BIOCHEMISTRY of PHOTOSYNTHESIS

BIOCHEMISTRY of PHOTOSYNTHESIS

R. P. F. Gregory

Department of Biological Chemistry,
The University of Manchester

WILEY-INTERSCIENCE
a division of John Wiley & Sons Ltd
LONDON · NEW YORK · SYDNEY · TORONTO

Library of Congress Catalog card No. 77-165955

ISBN 0 471 32675 5

Printed in Northern Ireland at The Universities Press, Belfast

Preface

Being asked to contribute an introductory remark reminded me that I have had the good fortune to witness the development of the biochemistry of photosynthesis almost from the time of its start. Yet the prejudices and polemics which attended the stages in this unfolding of knowledge no doubt left their scars on memory as the time continued to the present. Now, we have an account by a young author free from these retrospective effects. He has developed the subject partly from common knowledge and partly from an historical sense. Even the experts in their own more narrow vistas may be expected only to grumble softly. Easily intelligible for younger students and refreshing for the more elderly ones, Dr. Gregory has given a clear and balanced account which can be a foundation for further development in any chosen branch of this ever widening subject.

ROBERT HILL

Introduction

The process of photosynthesis, in which the energy of light (sunlight) is chemically captured by living organisms, and on which the whole of our planet's life-stock depends, is worthy of considerable stress in an undergraduate course of biochemistry, the more so if it provides for a synthesis of topics normally isolated by the necessarily linear nature of such courses. For example, the metabolic pathways of photosynthesis form a useful antithesis in the understanding, at an elementary level, of the pathways of glycolysis and the pentose cycle. The same applies to the cytochromes of the thylakoid, the production of oxygen from water, NADP reduction and photophosphorylation, all of which can be profitably compared with analogous processes in mitochondria. The chloroplast itself, in its relation to the cell, presents a most striking example of biochemical compartmentation. The first aim of this text is to provide an introduction to photosynthesis on the above basis. It is hoped that this introduction, Part I, will be with some selection valuable in courses of botany, and possibly at sixth-form level as well.

Since an introduction involves simplification, which is somewhat unsound, the text continues in Part II towards a second purpose, that of presenting an account of the subject giving the principal points of view (in 1970) and the experimental work and argument by which they were defended. As the writing progressed, however, it became clear that simplification would reappear, and the reader should be aware of two serious manifestations of it. First, the historical aspect and the development of the present-day concepts of photosynthesis, has been largely suppressed, except where old or obsolescent terms and hypotheses may still be unearthed by students and cause confusion. References in the text, and figures reproduced from previous publications, have in the main been selected for their utility and clarity of exposition rather than for evidence of priority. Secondly, the style adopted for the work is that of dividing up the field into a few areas, one to each chapter, and within each chapter to set out sections each written round a discrete point of view, aiming to cover the greater part of the area concerned.

In accord with the aim of supporting a course of lectures in this subject, I have given at the end of Part I a selection of problems, some numerical, some requiring discussion. To enable the most use to be made of Part II as a reference section, the index has been laid out in an extended manner, for which I am grateful to the publishers.

It is a pleasure to thank all my colleagues who have helped with suggestions, particularly Dr. I. West and Dr. A. G. Lowe. I am indebted to Mr. A. Green-wood, Dr. H. Bronwen Griffiths, Mr. R. Bronchart, Professor R. B. Park and Dr. G. Cohen-Bazire for electron micrographs, and Professor D. A. Walker for an autoradiograph. I gratefully acknowledge the painstaking criticism offered by Professor F. R. Whatley, Professor Walker and Dr. A. R. Crofts. My thanks are also due to Mrs. D. M. Warrior for typing the manuscript, and above all to my wife Julia for constant help and encouragement.

RICHARD GREGORY
Department of Biological Chemistry,
The University of Manchester.
Spring 1971.

Contents

PART 2

Part 1

CHAPTER 1

The context of photosynthesis

1.1 The energy of life

The biochemist investigates biological problems using the techniques of chemistry. Living matter is made up of materials which can be separated and analysed, and the changes that continually take place can be observed and even copied using extracted substances *in vitro*. One prominent goal is to explain the behaviour of living things in terms of identified chemical materials and their reactions. It is hard to think of any region of biology to which biochemists are foreign. Indeed it is through a biochemical approach that many biological topics may be seen to be related. In this text we shall examine photosynthesis, an activity mainly of green plants, and show how the problems raised by the phenomenon have been investigated, and how the importance of the ideas involved extends out of the world of plants and affects our understanding of the fundamental processes of life.

What processes of life would we regard as fundamental? Let us approach this by considering what, in chemical terms, a living organism is. First, we can recognize a discrete enclosed space, with a skin or membrane round it making a boundary. Inside the space is a structured system with solid and liquid phases containing proteins, lipids, carbohydrates, nucleic acids and other substances, all in an aqueous medium which is kept more or less constant in composition, regardless of wide variations in the external environment. As described so far, such a system could not be expected to persist for very long in the face of continuous disruptive processes. These include spontaneous and random hydrolyses of protein and nucleic acid components, and chemical reactions of many materials with oxygen. Also the boundary-structure, which must not be impermeable, tends to lose or gain water and solutes and thus destroy the internal environment of the organism. Again, the structures of the system are easily damaged by mechanical shock, heat and so on. Damage of this kind occurs continually and has to be made good. If the organism is to maintain its structure and size, and still more if it is to grow, it must have a source of energy available to it from the environment. Given this source of energy it can replace materials as fast as they are degraded, and given a source of component material as well, it can grow.

3

We therefore look for sources of energy available to living organisms. Only two are found, and these represent a clear dichotomy in biology. The first source is chemical energy. This is represented by a substance or mixture of substances that is unstable, meaning that an irreversible process is liable to occur to change the substance. When such a reaction takes place, we say that the system loses energy,* or that energy is released. The organism takes in the unstable material, and causes the reaction to proceed faster. This is catalysis,

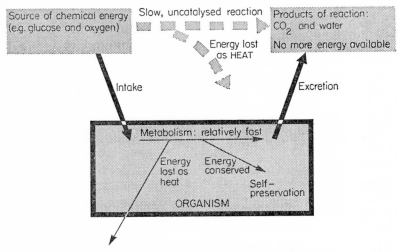

Figure 1.1. The energy supply of chemotrophic organisms

and the biological catalysts are termed enzymes. They are arranged so that the energy released by the reaction is partly conserved in a form which the organism can apply to the processes of repair and growth (see Figure 1.1). An example of a single substance providing energy by a catalysed breakdown is glucose undergoing fermentation to ethanol and carbon dioxide by yeast. Glucose and oxygen together yield considerably more energy in the chain of reactions leading to carbon dioxide and water. (There is an important chemical principle (Hess' Law) which states that the quantity of energy released by a process depends only on the initial and final states of the system, not on the nature of the intermediate pathways.) Most of the materials that yield chemical energy for life are in fact produced by living organisms, so that forms of life that require chemical energy (we shall term these *chemotrophs*,

---

* Usually heat is released. This is not always the case; if the system increases in disorder, it may proceed even if heat is absorbed. We use the term *free energy* to avoid the confusion. Free energy (G) combines heat (H) and disorder units (entropy) (S) in the relationship $\Delta G = \Delta H - T \Delta S$ (at constant temperatures).

and the life-style *chemotrophy*) consume the dead or living bodies of other organisms. It must be clear that as energy is lost to the environment, usually as heat, at each turn of the food cycle, so chemotrophic life by itself cannot last very long before a state of famine arises. Another source of energy is needed.

The second source is the energy of light. Of many other kinds of environmental energy source, thermal, gravitational, seismic, radioactive and electric, only electromagnetic radiation has been of any biochemical value,

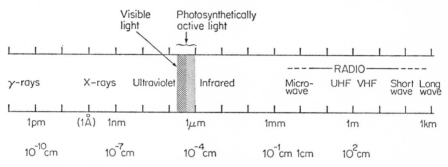

Figure 1.2. The range of biologically useful light in the spectrum of electromagnetic radiation

and of the immense range of such radiation, only that part known as visible light. The principle by which light interacts with matter and enables the organism to abstract its energy is interesting and instructive. Of the range of wavelengths of electromagnetic radiation, shown in Figure 1.2, those of the visible light band result in the formation of *electronically excited states* of certain types of substance (pigments). Longer wavelengths are strongly absorbed by water and very many biological materials, but give rise instead to the excitations detected as heat. Shorter wavelengths, on the other hand, do give rise to electronically excited states, but not only are these states so energetic that they cause random reactions harmful to the organism, but there are relatively many materials which absorb. Ultraviolet light is of no use to the organism, at least for the purposes of providing energy, and it appears that a screen of special pigments is often present that absorbs the harmful radiation safely.

Organisms that require light energy (termed *phototrophs*) form a dense screen of pigment that absorbs light at a wavelength long enough to avoid screening by other cell materials, and short enough for the energy to be trapped in an electronically excited state. This excited state is unstable; it will lose its energy by one of several processes. Although its lifetime is short, the phototroph acts to shorten it further, by catalysing its decay along a specific path, from which some of the energy can be conserved in a form which the organism can apply to the processes of repair and growth (see Figure 1.3). The similarity between chemotrophy and phototrophy is very marked. The biochemical process of obtaining energy from chemical sources is *respiration*, and that in which light energy is utilized is *photosynthesis*.

We have divided living organisms into the chemotrophs and phototrophs on the basis of the source of their energy. An alternative classification is that based on the nature of the nutrient that provides carbon material for the

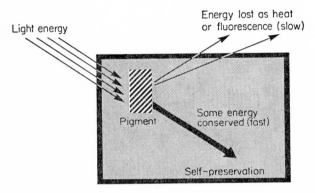

Figure 1.3. The energy supply of phototrophic organisms

growth of the organism. *Autotrophs* need only carbon dioxide, while *hetero-trophs* require organic substances. Earlier, autotrophic nutrition was considered to be synonymous with photosynthesis in plants, and heterotrophic nutrition with respiration in animals and bacteria. However the discoveries, first of the *chemosynthetic bacteria* which are able to grow on carbon dioxide, obtaining their energy from the oxidation of inorganic materials such as sulphur or ferrous iron, and secondly the *photosynthetic bacteria* which use light energy to grow on materials which may be either organic or inorganic, made this scheme of classification somewhat cumbersome. Thus the chemosynthetic bacteria were termed 'chemiautotrophic', and the two styles of photosynthetic bacteria 'photoheterotrophic' and 'photoautotrophic' respectively. As will be shown later in this text, several photosynthetic bacteria are able to practise either mode, depending on what substrates are available. All things considered, it seems easier and of more fundamental importance to make our distinction on the basis of the energy source rather than on the nature of the carbon supply. Table 1.1 shows how the two classes intersect.

Table 1.1. Classifications of organisms

		Carbon Source	
		CO_2 (Autotrophic)	Organic (Heterotrophic)
Energy Source	Chemical (Chemotrophic)	Chemosynthesis (bacteria)	Respiration
	Light (Phototrophic)	Photosynthesis (green plants and some bacteria) Photoautotrophic	Photosynthesis (bacteria) Photoheterotrophic

Figure 1.4 represents the passage of energy from light into the chemical form of the materials synthesized by the phototroph, and its subsequent utilization by the chemotroph. Heat is lost at every stage, so that in a steady-state biological situation, all the energy of light is eventually converted to environmental heat. The figure illustrates a further point: the chemical materials which make up the energy source for the chemotroph are in the main also the source of the material from which, with the energy, the

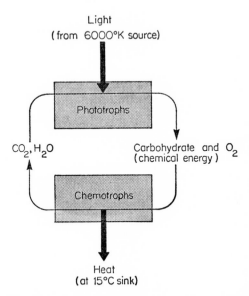

Figure 1.4. The dependence of chemotrophs on phototrophs, and of phototrophs on light from the sun, in terms of a flow of energy

chemotroph synthesizes new material and grows. In the case of the photo-troph, however, the nutrient material is often the stable waste materials of the chemotroph, such as carbon dioxide and water, from which no more chemical energy is available. In a steady-state system these nutrient cycles must balance. This balance is the basis of the familiar carbon-cycle diagram (see Figure 1.5). While this diagram shows small backwaters of the cycle,

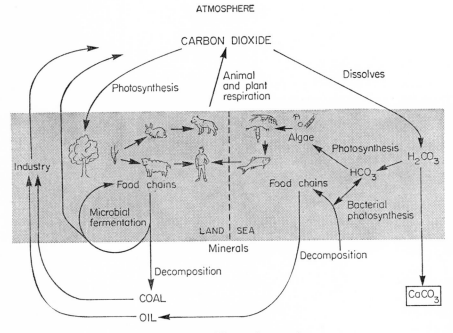

Figure 1.5. The carbon cycle

mostly involving bacterial fermentations and heterotrophic photosynthesis, it stresses that the processes of photosynthesis and respiration are closely interdependent. Even the processes carried on by the chemosynthetic bacteria are dependent, since their energy sources, oxygen, nitrates, ferrous iron and sulphur compounds are mainly produced by other organisms, the chief exception being the rare exposure of mineral sulphur.

The carbon cycle diagram, Figure 1.5, needs an indication of scale. Various estimates have been made of the annual quantity of carbon dioxide fixed by photosynthesis, the main difficulty being that the greater part takes place in the ocean. The ocean indeed contains most of the available carbon dioxide, although the atmosphere is the principal reservoir of oxygen. The mass of the atmosphere, 5.3×10^{18} kg, includes 1.1×10^{18} kg oxygen; the sea has been calculated to contain some 5.0×10^{16} kg carbon dioxide. Taking Riley's (1944) figure for the annual fixation of carbon dioxide as 5.7×10^{11} tons (7.0×10^{14} kg), the carbon dioxide of the 'biosphere' must be turned over every few hundred years. If we assume that the process of photosynthesis averaged out over the earth can be represented by the equation

$$CO_2 + H_2O \rightarrow O_2 + \tfrac{1}{6}C_6H_{12}O_6$$

we know that for each mole of carbon dioxide fixed a mole of oxygen is released. The annual rate of oxygen release corresponding to the above figures is 5.1×10^{14} kg,* which must turn the reservoir over in some 2000 years. These time scales are short; if the processes of respiration and photosynthesis were not in close balance, the composition of the atmosphere would have changed significantly even in so short an interval as historical time. We have to conclude that the dichotomies, with respect either to the carbon or energy sources of living organisms, are exact.

1.2 The nature of light

The nature of light has been a conceptual problem for many hundreds of years. Newton had regarded light as a stream of particles, a view which was later held to have been invalidated by the demonstration by Young of interference phenomena; rationalization of which was only possible on the basis of a wave theory, as shown in Figure 1.6. This view persisted, in spite of difficulties about the nature of the medium (the luminiferous ether) required to support the wave motion, until the beginning of the twentieth century. In 1905 Einstein showed that the photoelectric effect presented formidable problems. In the photoelectric effect, light incident on a suitable surface in a vacuum causes electrons to be emitted from it. The energy of the electrons is independent of the intensity of the light, which only affects their number, but is strongly dependent on the wavelength of the light (see Figure 1.7). Regardless of the intensity of the light, emission commenced with no perceptible time lag. It was necessary to conclude that all the energy incident on the surface as a whole could more or less instantaneously be gathered at a single atom and result in an electron emission, which made no sense on a simple wave theory. Secondly, the existence of a red limit on the wavelength of light for electron emission agreed with the theorizing of Planck five years previously; Planck had derived an expression for the wavelength distribution

---

* This figure 5.1×10^{14} kg is obtained by multiplying 7.0×10^{14} kg, the uptake of CO_2 by 32/44, the ratio of the molecular weights of O_2 and CO_2.

of radiation emerging from a cavity in terms of the temperature of the surrounding material. This expression, which agreed with experimental findings, was based on the novel assumption that energy in the atoms of the material was quantized, that is to say was only to take discrete values. In the transition from one energy state to another, a definite *quantum* of radiation

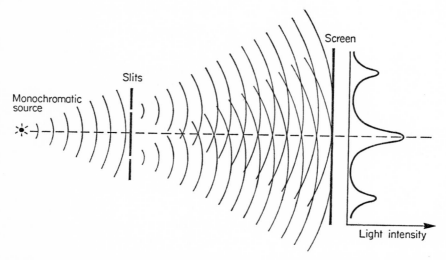

Figure 1.6. The phenomenon of interference in a beam of light passing through a double slit

energy was emitted, of which the wavelength could be calculated from the equations

$$E = h\nu \qquad \nu = c/\lambda$$

where E is the energy, ν the frequency, h a universal constant called Planck's constant, λ the wavelength and c the speed of light. In 1905 the photoelectric effect was explained in terms of quanta, particles or wave-packets, small enough to interact with atoms or even electrons, with energies governed by their wavelength. A certain energy was necessary to detach the electron, and additional energy increased the kinetic energy of the expelled electron, thus explaining Figure 1.7.

This view was further modified. It was found that beams of electrons displayed interference and diffraction behaviour. An electron appeared to possess a characteristic 'wavelength' depending on its energy. Yet electrons were thought to be material objects. This surprising result was rationalized by Heisenberg in the *Uncertainty Principle*, which sets a limit on the accuracy with which the position and momentum of a particle can simultaneously be known. This limit is so small that for particles as large as atomic nuclei, and of course all familiar material objects, it has no significance. However for light particles such as electrons and *photons* (a photon being the particle carrying a quantum of radiation) the area of uncertainty in the position of a particle of defined momentum is so great that it has to be expressed in terms

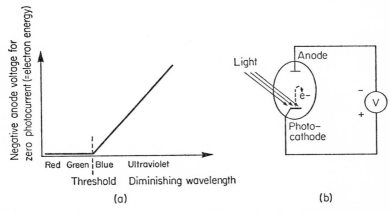

Figure 1.7. The photoelectric effect: (a) The existence of a threshold wavelength of light necessary to release electrons in the apparatus (b).

of statistical mechanics, or wave-mechanics, which yields a pattern. This pattern has the form of a wave, but the only interpretation of the wave-pattern is that the square of the amplitude at a given point is proportional to the probability of finding the electron there. If by means of an observation the position is actually determined, the wave pattern vanishes, and if the particle continues to move from that point onwards, it has an unknown momentum, and a new probability pattern is formed to describe its future appearance. Thus an electron or a photon allowed to fall on a surface with two slits as in Figure 1.6 has a probability of being found at points on the second surface given by the wave-pattern we know as interference fringes. In between the generation of the particle and its final detection, its position is not determined, and any attempt to find its path interferes with its momentum and ruins the experiment. This is apparently a fundamental uncertainty: there is no possibility of finding, for example, which slit it passed through in Figure 1.6. The usual attitude is to say that what can never be verified does not exist, or at least is of no concern to science. One could use the less exact phrase that electrons and photons propagate as waves but are detected as particles.

1.3 Photosynthetic structures

From the point of view of a chemist a living organism such as a spinach plant presents material that is not only heterogeneous but which also has an organization like a nest of chinese boxes; in pursuing the site of the photosynthetic process he will separate the leaves, then the mesophyll cells, then chloroplasts, and then proceed to disaggregate the chloroplast components. The further this is taken the greater the suspicion aroused in the mind of the biologist, who insists that the relations between parts are as important as the parts themselves. The fact remains however that in green plants the biochemical part of photosynthesis is almost entirely located inside the chloroplasts, which are convenient particles to prepare and investigate. Without implying that it is independent of its environment, we shall find the chloroplast and its components a useful system on which to base a biochemical survey of photosynthesis. Furthermore, although the structures and processes differ in the photosynthetic bacteria and the blue–green algae, which do not have chloroplasts, we shall find that it is convenient to study them in comparison with green plants rather than as an independent subject.

Plate 1 (facing p. 84) shows a section through a mesophyll cell of a broad bean leaf. The chloroplasts show up clearly as heavily-staining objects, which account for a considerable part of the protoplasmic mass. The layer of cytoplasm that covers the nucleus, chloroplasts and cell wall is very thin; most of the volume enclosed by the cell wall is the vacuole, a watery solution which may act as a waste-pool or a reservoir of water or solutes. This is typical of photosynthetic cells of higher plants. It is not surprising that the chloroplasts may contain up to 70% of the total cell protein. Chloroplasts vary in size and shape. Those of spinach tend to be some 5–7 μm (1 μm $= 10^{-6}$ m), and oblately spheroidal in shape. They tend to be distended by starch grains, or in algae by pyrenoids, which are temporary stores of starch and protein respectively.

The chloroplast illustrated in Plate 2 (facing p. 84) shows an external double membrane, each layer of which is a 'unit' membrane of the type found in many cell membranes such as the endoplasmic reticulum. This double membrane is termed the chloroplast envelope. (The term 'chloroplast membrane' can be confused with the lamellar system inside). The envelope is osmotically sensitive: if a leaf is cut with a razor in a drop of isotonic sucrose on a microscope slide, a few intact chloroplasts are released and may be observed under the microscope. Addition of a drop of water causes characteristic balloons to appear. Sometimes, when sections are observed under the electron microscope, 'blisters' are observed between the unit membranes forming the envelope. When chloroplasts are isolated from leaves by grinding and centrifugation, the envelope is easily damaged, and up to 1963–1965 this almost inevitably happened when the chloroplasts were washed free of cytoplasm. Damage causes loss of the *stroma*, which is the ground-substance staining moderately with electron-dense stains and which fills the chloroplasts and surrounds the lamellar structure. Such chloroplasts retain their general form, but there is some increase in size, and under the light microscope they lose their bright, refractive appearance and seem 'flat'; in most higher plants *grana* (spots of darker green) can be seen after this treatment.

The most striking feature of electron micrographs of chloroplasts is the system of lamellae. On inspection these turn out to be closely-flattened sacs termed thylakoids ('bag-objects'). The extent of the thylakoid system in any chloroplast varies both with the species and with the history of the individual plant. In many green plants including spinach there are many regions where the thylakoids are collected into stacks; these stacks are the *grana* referred to above. Individual thylakoids may continue from one granum to another, but the thylakoids in a granum (often known as discs, the granum being a stack of discs) appear to stain more densely or are thicker-walled than those outside the granum. Disruption of chloroplasts in a homogenizer or by ultrasonic treatment allows the thylakoid fragments to be collected, free from stroma and envelope. All the pigments of the photosynthetic system, chlorophyll and carotenoids, are present in the thylakoid membranes. The chloroplast is thus divided into three regions: the stroma, the 'solid' material of the thylakoid membranes, and the very small space inside the thylakoids. This separation of phases is important.

Also visible in the chloroplast are 'osmiophilic globules' (also termed plastoglobuli), so called because they take up the electron-dense stain osmium tetroxide, which contain lipid material. With special staining techniques, deoxyribonucleic acid (DNA) can be seen in the stroma. (DNA is principally found in cell nuclei, where it is part of the mechanism of Mendelian inheritance; it has been shown that the sequence of nucleotides that make up DNA governs the formation of at least some of the cell proteins.) The presence of DNA in the chloroplast raises interesting problems concerning the degree of independence of chloroplasts from nuclear genetic control. Under special conditions, ribosomes, the seat of protein synthesis in the cell, can be seen in the chloroplast as well. Both the DNA and the ribosomes of chloroplasts can be distinguished biochemically, in homogenates, from bulk of the DNA and ribosomes in the cell.

The blue–green algae (Cyanophyta) carry out an autotrophic photosynthetic process similar to that of green plants; they contain chlorophyll *a* as the principal pigment, and evolve oxygen. However there are some important differences: the thylakoids are not contained in chloroplasts, but appear to ramify throughout the cell (Plate 3 facing p. 84). The cytoplasm of the cell represents the stroma of the chloroplast of the higher plant. The algae show differences in their photosynthetic pigments: in green algae and higher plants chlorophyll *b* is an *accessory pigment* to chlorophyll *a*. In some other groups of algae chlorophyll *c* appears instead of chlorophyll *b*, and in others, including the blue–greens, there is no accessory chlorophyll but instead various biliproteins, which are formed by the combination of protein with various *bile pigments*, which while derived chemically from compounds of the chlorophyll type, are by no means similar in properties.

The photosynthetic bacteria present a very different picture, both in their organization and in the biochemical details of their photosynthetic process. They have been divided into the *purple* and the *green* bacteria, and the purple group has been divided further into *sulphur* and *non-sulphur* purple bacteria. However the distinctions between these groups are by no means clear cut. In general these bacteria have cell walls similar to those of other gram-negative bacteria, inside which is a thick cell membrane. This cell membrane often appears somewhat folded, and may give rise to membranes or vesicles within the body of the cell. When the cells are disrupted, these membranes may be obtained by centrifugation, and are termed *chromatophores* (not to be confused with chemical *chromophores*). The membrane system *in vivo* may or may not be recognizable as discrete chromatophores. Plates 4 to 7 facing p. 84 illustrate the membrane material in four species of photosynthetic bacteria.

The membranes of purple bacteria contain not chlorophyll *a* but bacterio-chlorophyll, which is closely related chemically to chlorophyll *a* but absorbs further into the red; *in vivo*, the pigments can make use of light well into the infrared region. The colour of the purple bacteria is due mainly to carotenoid pigments. Carotenoids (including the carotene and xanthophyll compounds) are universal in photosynthetic organisms, but they are not all necessarily involved in the capture of light for energy conversion in photosynthesis. Another role appears to be a protection in some way from the combined effects of oxygen and light, which damage living matter.

While in the green plants carbon dioxide is reduced using hydrogen from water (leaving oxygen to be evolved as a gas), in the photosynthetic bacteria this never occurs. Instead other hydrogen donors are employed; in the green and the purple sulphur bacteria sulphide or thiosulphate is used, and in the purple non-sulphur bacteria a variety of organic compounds may be oxidized and the hydrogen used for reduction of the carbon source. The carbon source may be carbon dioxide (autotrophic nutrition) or an organic compound (heterotrophic nutrition) which may even be the same material as was used for the hydrogen donor. It is remarkable that in at least one group, the purple non-sulphur bacteria, metabolism can switch to an aerobic, chemotrophic growth, and the formation of pigment stops until such time as photosynthesis is required again. For the characteristic features of photosynthesis, we must therefore look first not at the metabolic reactions of the hydrogen donor and carbon substrates, but rather at the unique energy conversion and conservation system.

References

Riley, G. A. (1944). The carbon metabolism and photosynthetic efficiency of the earth as a whole. *Amer. Scientist*, **32**, 129.

Suggested further reading:

Fogg, G. E. (1968). *Photosynthesis*, Edinburgh University Press, London, Chapter 7.
Rabinowitch, E. and Govindjee (1969). *Photosynthesis*, Wiley, New York, Chapters 1–4.

CHAPTER 2

The absorption of light

2.1 Pigments and chromophores

A pigment is a chemical substance that absorbs visible light. That part of the molecule which is responsible for the absorption is termed the *chromophore*. In small molecules the chromophore may be inseparable from the whole structure, just as all the atoms of say formic acid contribute to the observed acidic properties. In larger molecules however certain groups can be recognized as chromophores regardless of the rest of the molecule, just as in general the group ·CO·OH can be recognized as a structure conferring acidic properties. Obviously this is only a matter of convenience, since the exact properties of chromophores (and the dissociation constants of acids) are bound to be affected by the other atoms of the molecule.

In biological materials there are relatively few types of molecular structure that absorb visible light. In section 1.1 it was pointed out that absorption in this region of the spectrum was associated with the excitation of specific electrons of the molecule to excited states; excited states readily underwent specific chemical reactions providing the basis of photosynthesis and vision. In the infrared region the energy of the quanta absorbed by biological systems (which are aqueous) is directly converted to thermal energy. Some bacteria can utilize light of wavelength up to 900 nm or more for photosynthesis, which indicates rather that the upper limit of visible light, or the lower limit of the energy of electronically excited states, is not precisely defined, rather than any exception to the rule above. In the ultraviolet region, with the increasing energies of the promoted electrons, the rate of reaction of the activated molecules is likely to be substantial even without catalysis. Furthermore, the number of possible electron transitions in a molecule increases with decreasing wavelength. Hence not only is absorption of ultraviolet light more widespread in biological materials, but the absorption tends to be associated with diverse and uncontrolled reactions.

2.2 Excited states

We will now turn to examination of the process of light absorption in more detail. It must be borne in mind however that the quantitative aspects of this account can only be derived for the simplest molecules, and the pigments that are important in light absorption for photosynthesis are not at all simple. More rigorous accounts of the processes involved can be found in Thomas (1965) or in Seliger and McElroy (1965).

A molecule contains many electrons. Some are 'inner' electrons, such as the $1s$ electrons of carbon, and for our purposes can be ignored. Other 'inner' electrons are the d-electrons of transition elements, and these will be discussed later as a special case. Of the 'outer', or valence electrons, some will be 'non-bonding' ('n-electrons'): they will be centred around one atom, and their behaviour will be only slightly perturbed by the presence of the

Figure 2.1. To illustrate the symmetries of electron patterns constituting σ- and π-bonds.

other centres. Electrons which form bonds are usually electrically centred around two atomic nuclei. If the electron distribution is cylindrically symmetrical around the axis of the bond, it is said to be a sigma (σ) bond; if not, it is said to be a pi (π) bond. In the 'double bond' of, for example, ethylene ($CH_2{=}CH_2$) one bond is π, the other σ. Diagrams are given to illustrate these bond types (see Figure 2.1) but it should be stressed that they do not illustrate the situation in any particular case. Electrons are restricted to orbitals in bonding just as in non-bonding cases. Each orbital type has its family of excited states. In thermal equilibrium at room temperature, a population of molecules has a large majority of electrons in the lowest, or 'ground' states of their respective orbitals. The excited states can be written (σ^*), (π^*) for (σ), (π) orbitals respectively. We can therefore write out a variety of types of electronic transition: $\sigma \rightarrow \sigma^*$, $\pi \rightarrow \pi^*$, $n \rightarrow \sigma^*$, $n \rightarrow \pi^*$. It happens, however, that the energy gap between the ground level and the first excited state of virtually all σ-bond electrons is so great (of the order of 10 eV) that the wavelength of the related photon lies in the far ultraviolet (124 nm for 10 eV). So far as biological systems are concerned, we may confine our attention to the $\pi \rightarrow \pi^*$ and $n \rightarrow \pi^*$ transitions, which may have energies corresponding to the range of visible light.

Figure 2.2 relates the two principal absorption bands of chlorophyll to the formation of the 'first' and 'second' excited states of the molecule. Although the energy stored by the second excited state is much greater than that in the first, the excess is not available for photosynthesis: the second excited state loses its energy in a transition to the first excited state, with the evolution of heat, in an extremely short time. Thus the result of the absorption of a 'blue' quantum is the same excited state as would have been produced by a 'red' quantum directly.

The figure also indicates the manner in which the ground state and the excited states each possess a series of *vibrational substates*. At biological

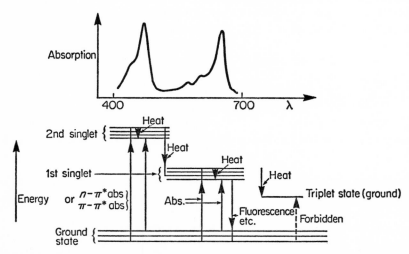

Figure 2.2. The relationship of the principal absorption peaks of chlorophyll with the excited states of the molecule. (The vibrational substates are represented impressionistically)

temperatures there are several substates populated at each level, so that there is a range of electronic transitions with slightly different energies. This is one of the reasons for the observed absorption bands being broad; at liquid nitrogen temperatures (77°K) spectra tend to be sharper. Another broadening influence is the tendency of chlorophyll molecules to aggregate in solution, and even more so in the chloroplast lamella. The absorption bands are broader and displaced to the red, and represent transitions of the aggregate rather than of individual chlorophyll molecules.

It may be appropriate to mention here the transitions of the *d*-electrons of transition metals. In the atoms of these elements *d–d* transitions are 'forbidden', that is to say take place relatively rarely, so that the light absorption is not very intense. However in complexes of the metals it may happen that an electron may pass from an orbital of one atom to one on another. This is termed a charge-transfer absorption, and is intense. A good example is the intense absorption of red light operating the transition of an electron between ferro–ferric iron atom pairs in Prussian blue (ferrous ferricyanide). In the chloroplast there is a copper protein, plastocyanin, which also shows an intense absorption (relative to say copper sulphate, atom for atom of copper) which may be due to such a transition. Charge-transfer excitations of transition-metal elements are not, however, likely to play much part in photosynthesis, because the absorption by other pigments is many orders of magnitude greater in the chloroplast. Apart from these *d–d* transitions, charge transfer complexes also occur in organic systems and may have a secondary importance in photosynthesis. For the present it would appear that $n \rightarrow \pi^*$ or $\pi \rightarrow \pi^*$ transitions of chlorophyll are the basis of light absorption in the chloroplast.

2.21 The triplet state

An orbital can contain up to two electrons, which must then have 'anti-parallel' spins. This may be visualized as in Figure 2.3 where the arrows

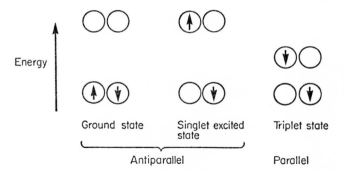

Figure 2.3. The formation of the triplet state preserves the resultant spin value, but reversal takes place when the triplet is formed

represent the magnetic axis resulting from the rotation of the electronic charge.

The two directions are arbitrarily given *spin quantum numbers* $+\frac{1}{2}$ and $-\frac{1}{2}$. When one of the electrons is promoted to a higher orbital, by absorption of a quantum of radiation or by other means, the direction of spin of the electron is not normally changed. The algebraic sum of the spin quantum numbers of all the electrons of the molecule, known as the *resultant spin, S,* is unchanged; in the case of chlorophyll in which all electrons are paired, $S = 0$. However there is a small probability that spin reversal may take place during the lifetime of the excited state (see Figure 2.3). When this happens, S becomes 1 (arbitrarily considered positive). This state once formed has a similarly low probability of spin re-reversal so that the lifetime of this new state is longer, and can often be recognized by spectroscopists. The spectroscopic terms for the two types of excited state are 'singlet' and 'triplet' respectively, referring to the *spin multiplicity* $(2S + 1)$. The triplet state may be thought of as the ground state of a new family of states: the first excited triplet, the second, and so on. This is the basis of the spectroscopic observation; the triplet is a new molecule and has a unique spectrum. Chemically, the triplet is more reactive, acting as a *bi-radical*, that is, as it has an unpaired electron in each of two orbitals, it is a free radical twice over. The triplet state is believed to be responsible for most photochemical reactions of dyes in solution, and a hypothesis by Franck and Rosenberg suggested triplet-chlorophyll to be a key intermediate in photosynthesis (see section 8.23). It must be stressed however that because spin reversal is an improbable process, the concentration of triplets remains low even in strong light; in the case of chlorophyll the triplet is hard to detect even in solution, and *in vivo* it has not been shown directly to exist.

2.22 The conversion of the energy of excitation

Fluorescence

The simplest degradation of the first excited singlet state is the reversal of the absorption. The electron falls back to the ground state and a photon is emitted, a process known as fluorescence. In a molecule some of the energy is lost as heat because the vibrations and rotations of thermal energy broaden the energy gap; on absorption of light an electron is in fact promoted to some vibrationally excited sub-state of the electronically excited state. The vibrational excitation is usually lost in 10^{-12} s, while fluorescence takes some 10^{-9} s. Light emitted by fluorescence from a molecule at room temperature is somewhat redder than that which excited it, the difference representing the lost vibrational energy. This is known as the Stokes' shift.

Internal conversion

The excited state of a molecule is not necessarily stable, and the energy may be degraded to thermal energy in vibrational, rotational and (ultimately) translational modes.

Both fluorescence and internal conversions typically take place in about 10^{-9} s. On the other hand if the triplet state is formed the lifetime is extended typically to 10^{-5} s. (Some compounds are known in which triplet states persist for several minutes.) Light emission from the triplet state, which has to be accompanied by spin reversal, is known as phosphorescence and has a longer wavelength than the normal fluorescence for the particular system. Alternatively the triplet can be converted back to the singlet, a process with a low probability. Nevertheless, this requires only a small amount of energy and can be achieved by thermal excitation, and 'delayed' fluorescence with the normal fluorescence spectrum can result.

Collisional deactivation

The activation energy is released as heat by means of a collision with a second molecule. In aqueous solutions at normal temperatures the collision frequency might be of the order of 10^{11} s^{-1}. In the thylakoid membrane, with chlorophyll molecules in a semi-solid state, a lower frequency of collisional deactivation might be assumed.

Energy transfer or migration

An excited molecule can return without collision to the ground state while a second molecule is raised to the excited state. When the molecules are a considerable distance apart (of the order of the wavelength of the light) the mechanism for this process is likely to be fluorescence followed by reabsorption, an inefficient process. In more concentrated solution there is a mechanism, known as *inductive resonance*, in which there is no intermediate photon, the receiver responding to the electric field of the transmitter. This process is far more efficient. The thylakoid may perhaps be regarded as a 'solution' phase of chlorophyll, of very high concentration, or it may perhaps be thought of as a solid. The topic of energy migration through a solid is interesting and is discussed on its own in section 2.4.

Chemical change

The third means by which an excited molecule can lose its energy is by taking part in a chemical reaction. The pigment molecule may itself undergo change—we are familiar with dyed fibres bleaching in sunlight—or it may not; so far as we can see, the latter is the case in photosynthesis. Again, the light energy may provide only the 'activation energy' required to bring the two reactants together, or, as in the case of photosynthesis, be partly conserved in the chemical energy of the products. Several types of reaction are possible.

Dissociation:

$$A—B \xrightarrow{h\nu} A + B$$

The energy of visible light quanta is in the region 40–80 kcal per einstein. (The einstein is 6×10^{23} (Avogadro's number) photons, i.e. a 'quantum mole'.) This does not normally permit rupture of a covalent bond. One example however is the dissociation of the ferricyanide ion by blue light.

2

Ionization: The excited molecule may lose an electron:

$$AB \xrightarrow{h\nu} AB^+ + e^-$$

Since the electron must be accepted by another molecule, this is an oxidation–reduction process. The molecule may alternatively dissociate into positive and negative ions:

$$A\!-\!B \xrightarrow{h\nu} A^+ + B^-$$

Thirdly, in a solid state matrix such as a crystal or possibly in the thylakoid membrane, an electron may be expelled which can migrate through the matrix with an extended lifetime. This is the basis of photoconductivity, and will be discussed later.

Association: The excited molecule can combine or otherwise react with a second (unexcited) molecule:

$$A^* + A \rightarrow A_2$$

In a fourth mechanism, *photosensitization*, the excited molecule can sensitize a reaction without appearing to undergo even temporary chemical change. A typical case is the dissociation of hydrogen molecules into atoms, sensitized by mercury vapour:

$$Hg \xrightarrow[253 \cdot 7 \text{ nm}]{h\nu} Hg^*$$

$$H_2 + Hg^* \longrightarrow 2H + Hg$$

2.3 The time scale

Kamen (1963) arranged the events of photosynthesis diagrammatically so as to emphasize the time scale on which the absorption of light and successive phenomena took place. He used the symbol pt_s by analogy with pH for the negative logarithm of the time in seconds. Thus the time taken to promote an electron being 10^{-15} s, the promotion is said to occur at pt_s 15. Figure 2.4 sets out events on such a logarithmic timescale. This approach, which may be unfamiliar, provides a new perspective on photosynthesis, and illustrates in a novel way the contributions made to the subject by various scientific disciplines.

2.4 The solid state: migration of energy and photoconductivity

The process described on p. 20, in which energy was passed from an excited molecule to another, can take place in various ways. The simplest case would be that in which fluorescence took place followed by reabsorption in the second molecule. Where distances are short (of the order of 6 nm) however, the process can take place without an intermediary photon. It has been described in detail only for crystals, usually crystals of elements, and the application to non-crystalline systems such as chlorophyll in thylakoid membranes is theoretically difficult.

In crystals two migrations can be distinguished. The first is the *fast migration* and depends on the regularity of the crystal: since all points tend

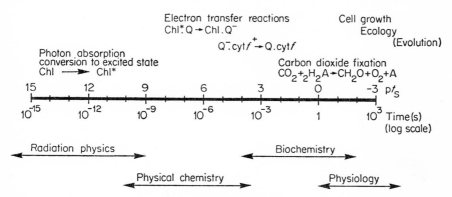

Figure 2.4. The 'pt_s' scale for describing biological activity, after Kamen (1963). The pt_s value is the negative logarithm of the time taken by a given process. From Kamen, M. D. (1963), *Primary Processes in Photosynthesis*, Academic Press, New York, p. 4, with permission. Copyright held by Academic Press.

to be identical, the excitation can be considered to belong to the crystal as a whole. After the absorption of the photon the energy is located at one lattice-point; as time elapses it migrates so that the number of points at which it might be found increases, according to the uncertainty principle. The quantum theory still applies to energy in a crystal as above, and the unit is the '*exciton*'. One quantum generates one exciton.

The *slow migration* is to be visualized as a process involving electron–'hole' pairs. The excited electron is promoted to an orbital which is large enough to cover several lattice-points. It can exchange with an electron of another molecule. By this exchange the electron changes its centre of orbit, which leaves behind a net unit of positive charge. This is termed a 'hole' and is analogous to a bubble in water. As the exchanges continue the excited electron moves through the crystal, and the hole follows it. This electron–hole pair is also termed an *exciton*. The average time for one jump has been estimated as 10^{-12} s, so that each exciton may make over 1000 jumps before it decays.

Imperfections in a crystal such as 'lattice defects' or the inclusion of alien material may form *traps* for either electrons or holes. When this happens the other carrier is free to move independently, so that semiconductivity results. If the electrons originated from pairs generated by light, the material is a photoconductor. It is possible for a heterogeneous crystal system to separate charges at points on its surface, under the influence of light.

There is considerable relevance to the photosynthetic systems here. Some indications of photoconductivity have been found in microcrystalline systems of the chloroplast pigments, and to a lesser extent in preparations of dried thylakoids and bacterial chromatophores themselves.

2.5 The pigments of photosynthetic systems

Of the various materials that absorb visible light and are found in photosynthetic structures, not all are to be regarded as photochemical sensitizing agents. Thus the protein materials, flavoproteins, plastocyanin (which contains copper), three or more cytochromes (protein–haem complexes) and ferredoxin (a protein containing iron and sulphide ions) and numerous quinone–quinol systems, appear to have more probable roles as electron-transport agents than as photochemical agents, and also absorb only a small part of the total light absorbed by the photosynthetic structure.

Other pigments are found in higher concentrations, and do not have any obvious chemical role. The most convincing evidence that these are the pigments responsible for trapping the energy of light comes from comparison of the absorption spectra with the action spectrum (see Figure 2.6) for photosynthesis. From such a comparison it seems certain that chlorophyll *a*, carotenoids and certain other pigments are the primary absorbers. The pigments in these classes are described below.

2.51 Chlorophyll

The chlorophylls are highly coloured substances, appearing virtually black in the solid condition, and green in solution. They are insoluble in water but dissolve readily in polar organic solvents. Chemically they are neutral in solution, and possess several recognizable groups. The most obvious of these is the square ring of rings, the tetrapyrrole system of the porphyrin type. This ring system is common to the haem pigment of blood and the cytochromes. Metal ions are often found complexed in the tetrapyrrole ring: in haem it is iron, but in chlorophyll magnesium. Other examples are copper in the red pigment of some birds' feathers (e.g. *Turaco*) and cobalt in vitamin B_{12}. There are two carboxyl groups in chlorophyll, but both are esterified, one with methanol and one with long-chain alcohol *phytol*. Phytol is an isoprenoid; one of the distinguishing features of isoprenoids is the arrangement of side-chains and the $5N$ number of carbon atoms (here 20, $N = 4$). This is because the molecule is built up of five-carbon isopentenyl units thus:

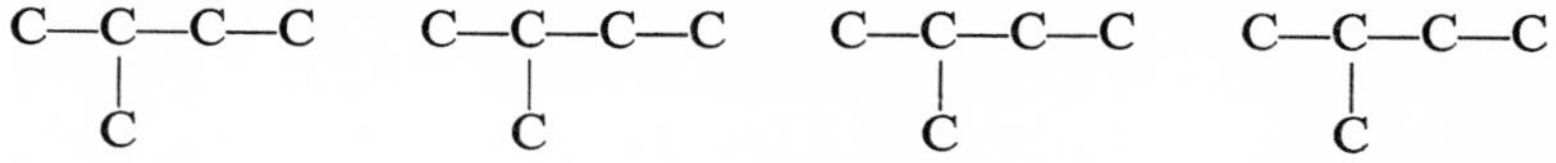

Other isoprenoids include steroids, carotenoids, natural rubber and the side-chains of the many quinones of biochemical importance. Many of these are involved in photosynthesis.

For an account of the isolation procedure and chemistry of chlorophyll the reader is referred to Hill (1963); only a few details can be given here. Reference to Figure 2.5, in which the formulae of several forms of chlorophyll appear, shows that their structure is related to porphyrins, in particular, proto-porphyrin IX. Magnesium occupies the centre of the square ring by displacing the acidic hydrogens. All four magnesium–nitrogen bonds are more or less equivalent, despite the conventional formula; the tetrapyrrole ring is fully conjugated so that all the bonds have some double-bond character. In the

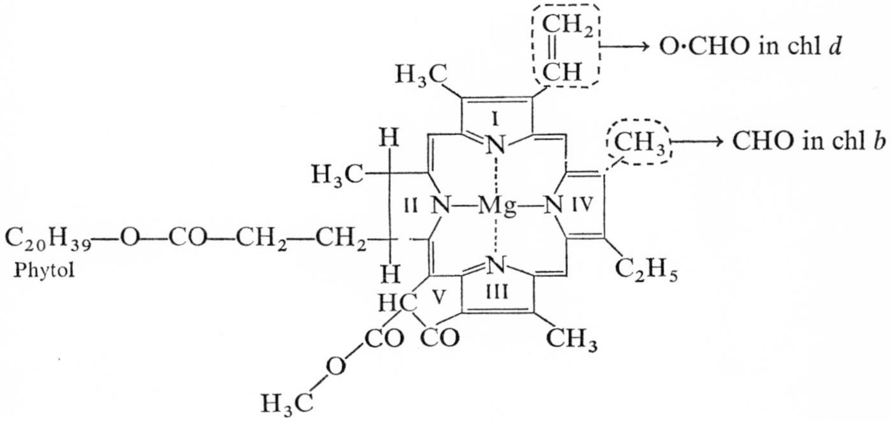

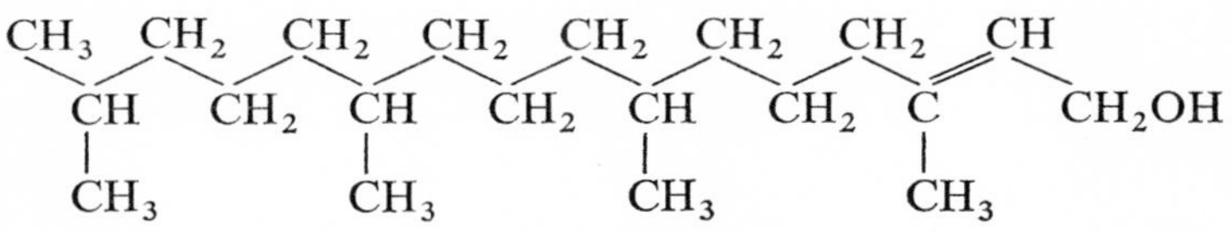

Phytol

I, II, III Chlorophyll *a, b, d,*

Figure 2.5. Formulae of the principal photosynthetic pigments

Figure 2.5 (*contd.*)

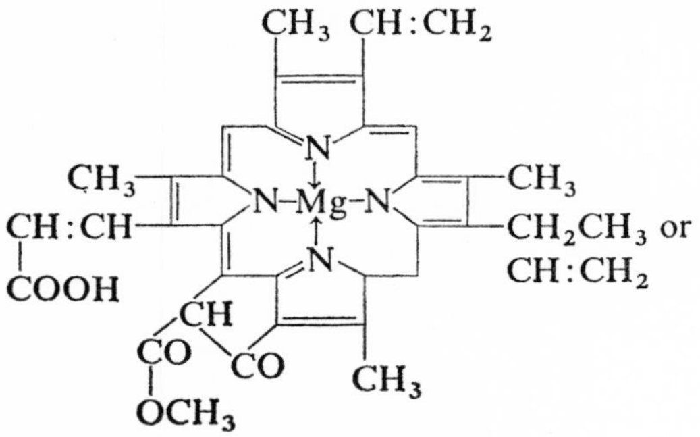

IV Chlorophyll *c*

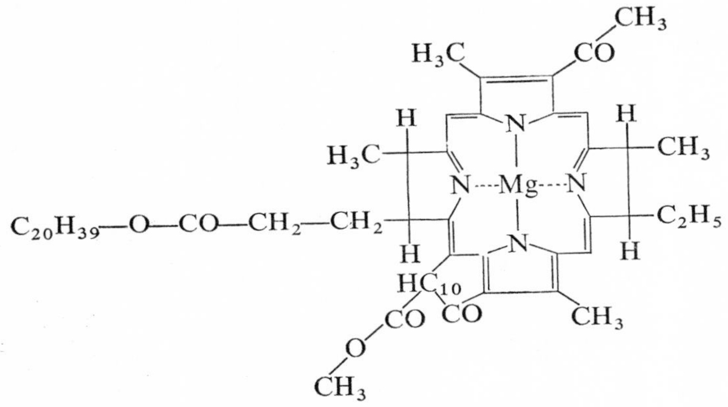

V Bacteriochlorophyll (*a*)

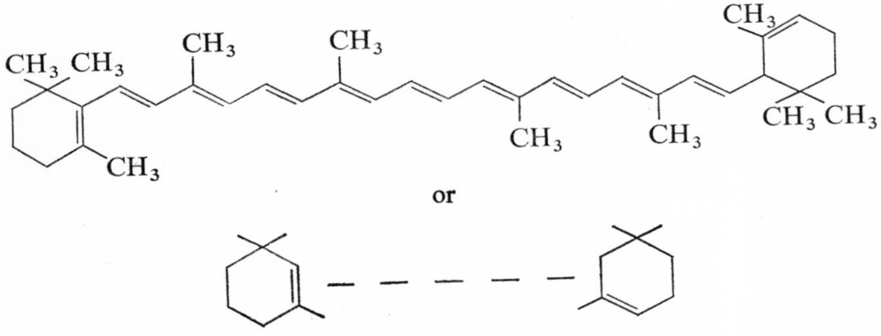

or

VI α-Carotene

Figure 2.5 (*contd.*)

VII *β*-Carotene

VIII *γ*-Carotene

IX Neoxanthin (uncertain)

X Violaxanthin

XI Lutein

XII Antheraxanthin

XIII Zeaxanthin

Figure 2.5 (*contd.*)

XIV Spirilloxanthin

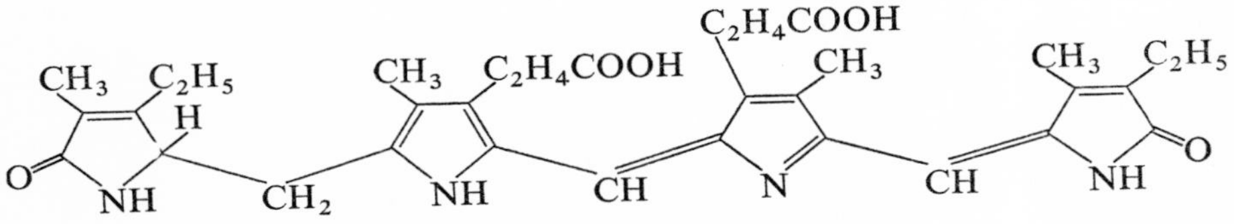

XV Probable formula for phycoerythrobilin

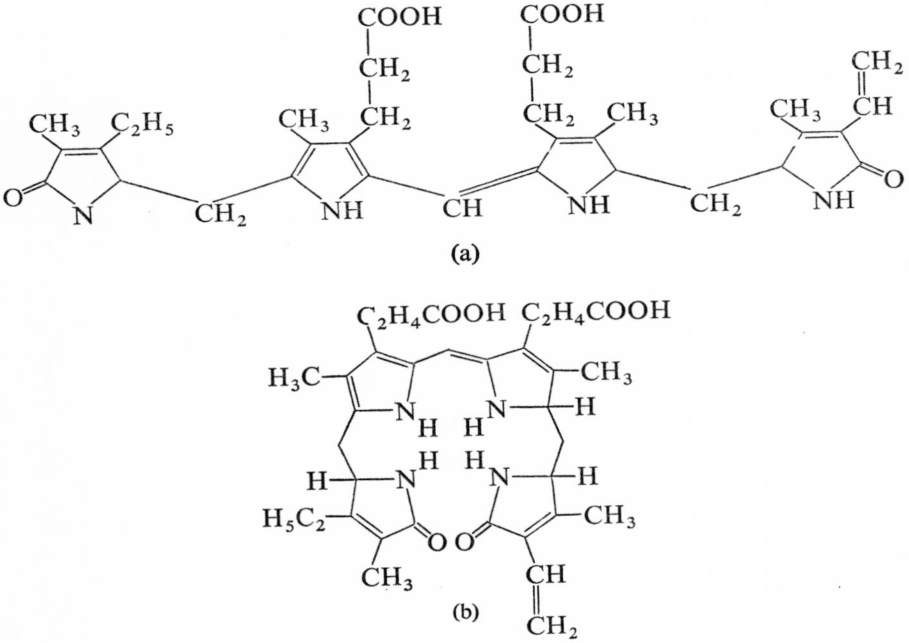

(a)

(b)

XVI Linear (a) and helical (b) formula for *d*-urobilin (see O'hEocha, 1966)

chlorophyll series, ring IV is saturated at positions 7 and 8, and in bacterio-chlorophyll, ring II undergoes a similar saturation as well. The effect of these modifications is to displace the porphyrin absorption spectrum to the red.

Ring V is peculiar to the chlorophylls. Carbon-10 carries an acidic hydrogen atom which may be important in the photochemical reaction. There are two carboxylic acid groups, one esterified with methanol and the other with the isoprenoid alcohol phytol. Hydrolysis of the latter by the specific enzyme *chlorophyllase* (a chloroplast enzyme that only appears to act in strong aqueous organic solvents) yields *chlorophyllide*. The reaction in ethanol or methanol yields the ethyl or methyl chlorophyllide. Hydrolysis of both esters by alcoholic potash gives *chlorophyllin*. Both chlorophyllide and chlorophyllin are soluble in millinormal alkalis. Loss of magnesium from chlorophyll, which is rapid in acid solution, forms *phaeophytin* from which *phaeophorbides* are obtained by hydrolysis of the esters. In general, the chlorophylls are derivatives of dihydroporphin or tetrahydroporphin; these derivatives are termed *chlorins* when the cyclopentanone ring (ring V, the isocyclic ring) is absent, *phorbins* if it is present. *Allomerized* chlorophyll is a heterogeneous product formed by oxidation of chlorophyll in solution in air. Oxidation takes place at position 10, in ring V.

Chlorophyll a has the formula I in Figure 2.5. It is universal except in the bacteria, and is usually the major pigment component of the thylakoids, by weight. From the absorption spectra in Figure 2.6, it can be seen that there are two principal peaks, in the blue and red regions. The absorption maximum in 80% acetone, the most convenient solvent for spectrophotometric estimation of the pigment, is 663 nm; in less polar solvents it is shifted to the red, and in the virtual absence of traces of water can be found at values of around 680 nm. Aggregates of chlorophyll *a* in concentrated solutions show a small shift to red, but crystalline specimens may show a maximum at 740 nm. In the thylakoid membrane, the absorption peak of chlorophyll *a* is much broader than in extracts of the same tissue in a solvent. This has been explained by supposing that chlorophyll exists in possibly five different types of environment, or aggregated states, such as might occur at different sites in a lipoprotein membrane structure. These five or more chlorophyll 'forms' are distributed between 664 and 703 nm, the principal component absorbing at 678 nm. It should be stressed that there is no reason to believe that these absorption maxima are caused by any other pigment than chlorophyll *a*, and that the differences are physical, not chemical.

Chlorophyll b has the formula II in Figure 2.5. The apparently slight difference in the formula is associated with a marked difference in the absorption spectrum (Figure 2.6) compared with chlorophyll *a*; the band in the red

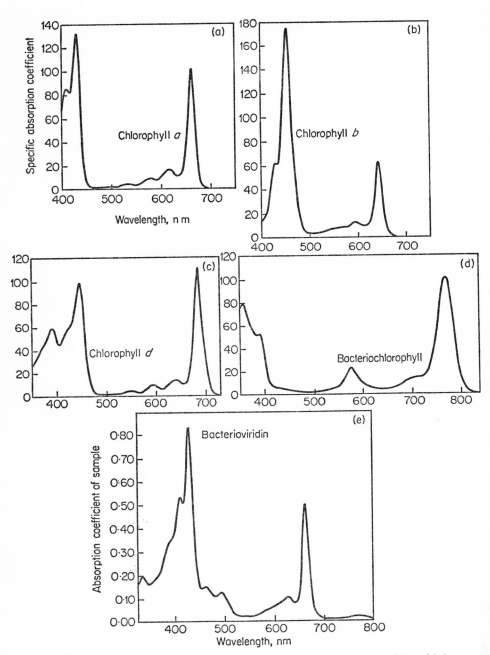

Figure 2.6. Absorption spectra of chlorophylls and other pigments. (a) to (e) from J. H. Smith and A. Benitez (1955), *Moderne Methoden der Pflanzenanalyse*, Vol. IV, Berlin–Göttingen–Heidelberg: Springer, p. 142–196, with permission. (f) from T. Tanada (1951), *Am. J. Bot.*, **38**, 276, with permission

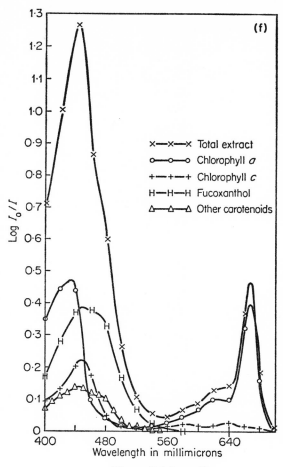

Figure 2.6 (*contd.*)

occurs at 635 nm in 80% acetone, and is half the intensity, while the blue-absorbing peak is slightly higher and shifted to the red. Chlorophyll *b* accompanies chlorophyll *a* in green algae and higher plants, except for a few individual species and mutant forms.

Chlorophyll c is found with chlorophyll *a* in diatoms (Bacillariophyceae), brown algae (Phaeophyta) and dinoflagellates (Pyrrophyta). It appears to be a mixture of two compounds whose formulae are given in Figure 2.5 (IV). (Dougherty, R. C. and coworkers (1970).) Ring V is present but phytol is absent, and the pigment is only soluble in polar organic solvents.

Chlorophyll d probably has the formula III (Figure 2.5). It occurs with chlorophyll *a* in some red algae (Rhodophyta).

Other green pigments, probably variants on the chlorophyll theme, are found in other algal groups.

Bacteriochlorophyll (V) differs from chlorophyll *a* in the 3,4 saturation, and also in the side chain at position 2. It is found in the purple bacteria, both Thiorhodaceae and Athiorhodaceae. In the chromatophores of these families the bacteriochlorophyll absorbs at much longer wavelengths, 800, 850 and 890 nm, compared to the pigment extracted into organic solvents which absorbs at approximately 770 nm. A second form of bacteriochlorophyll is known, bacteriochlorophyll *b*, but only in one species; it absorbs at 795 nm, (960 nm in the cells).

Chlorobium *chlorophyll-660* is the principal pigment of the third group of photosynthetic bacteria, the green sulphur bacteria (Chlorobacteriaceae). This pigment was originally termed bacterioviridin. The formula is uncertain; although Fischer and Stern (1940) suggested that it was 2-acetyl chlorophyll *a*, there appears to be no methoxyl group; an ester of an alcohol other than phytol seems to be present, and the phase test, which depends on the hydrogen atom on carbon-10, is negative. If the cyclopentanone ring is absent it may mean a reappraisal of the role of this part of the molecular structure in the photochemistry of chlorophyll in general. Shifts to the red occur in the green bacteria also, the *in vivo* absorption at 740 nm altering to 665 nm on extraction of the pigment.

2.52 Carotenoids

The *carotenes* are hydrocarbons, formed by the isoprene pathway of biosynthesis, which is revealed by the carbon skeleton being a multiple of a five-carbon unit with a characteristic branching pattern. The majority are C_{40} compounds.

α-Carotene (VI) is a minor carotene component, except in the Siphonales (green algae); in all other green plants *β-carotene* (VII) is the major component. The green sulphur bacteria, on the other hand, have *γ-carotene* (VIII). Only traces of other carotenes are known to exist in photosynthetic systems. The carotenes are soluble in petroleum ether, in contrast to chlorophyll *a* which requires traces of polar solvent to be present. The xanthophylls while soluble in petrol ether are readily extracted by 90% methanol solution. A possible anti-oxidant role for some carotene pigments is discussed in a later chapter.

The *xanthophylls* are oxygenated carotenes, and are of a great variety. Their numbers exceed the total of chlorophylls and carotenes together, and thus only a few can be described here.

The green algae and higher plants contain principally neoxanthin (IX), violaxanthin (X) and lutein (XI). To these we may add antheraxanthin (XII) from the Euglenales, and fucoxanthin and zeaxanthin (XIII) from the brown algae. Spirilloxanthin (XIV) is the major xanthophyll in the purple non-sulphur bacterium *Rhodospirillum rubrum*.

The xanthophylls undergo interconversion during photosynthesis, at least in green plants that evolve oxygen. These pigments may protect the thylakoids from damage by excess light and oxygen. All the plant pigments are sensitive to light and oxygen when isolated, necessitating careful chromatographic procedure if confusing artifacts are not to be so produced. Strain (1966) has given a useful review of these methods.

2.53 Biliproteins

Bile pigments are linear tetrapyrroles formed by the opening of the porphyrin ring. Biliproteins are proteins conjugated with a bile pigment prosthetic group. In plants, biliproteins are confined to the Rhodophyta, Cyanophyta and Cryptophyta. This is to exclude phytochrome, which although apparently widely distributed is not known to have any role in photosynthesis, but rather in the field of photoperiodic phenomena. There is evidence that the algal biliproteins are aggregated on the outsides of the thylakoids (see Plate 8, facing p. 84) giving a granular appearance. They are in general soluble, and tend to leach out during isolation of the photosynthetic organelles.

The red biliproteins, the phycoerythrins, are classified into the R, B, C and cryptomonad phycoerythrins, originally indicating their principal source. The blue pigments are the phycocyanins, with the prefixes R, C, allo and cryptomonad. O'hEocha (1966) gives a useful review of the known distribution among the algal groups. The prosthetic groups are phycoerythrobilin (XV), phycocyanobilin and phycourobilin. A helical structure (XVI) is almost certainly more appropriate than the linear scheme in all cases.

The view taken in this text, at least for the purpose of discussion, is that chlorophyll *a* in green plants and the pigments absorbing furthest into the red in the various bacteria are the primary pigments, and the carotenes, xanthophylls and biliproteins as well as the other chlorophyll varieties together are classed as *accessory pigments*. Evidence for this view will be given later.

2.6 Primary and accessory pigments

The basic theory for the operation of accessory pigments in photosynthesis is based on the observation that chlorophyll *a* (or bacteriochlorophyll) is the pigment absorbing furthest into the red; that is, that has the least energy of excitation. It is energetically feasible, therefore, for excited states of the other pigments to pass their energy to chlorophyll *a* by the process described in section 2.4, or alternatively for molecules of chlorophyll *a* to act as 'sinks' or 'traps' for some form of exciton or electron migrating from the site of the absorption of a photon. For this to be possible of course the pigments must be in suitable physical proximity. Secondly, for energy migration to occur,

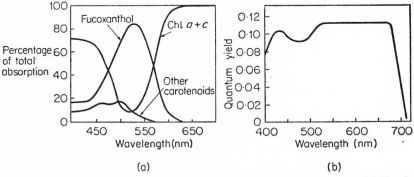

(a)   (b)

Figure 2.7. The operation of accessory pigments in *Navicula minima*. (a) To show that the light absorbed by the cells is distributed among the pigments, the percentages varying with the wavelengths. (b) To show that the quantum yield is relatively constant with wavelength, and hence that energy is being passed to chlorophyll by pigments such as fucoxanthol. From T. Tanada (1951), *Amer. J. Bot.*, **38**, 276, with permission

the absorption peak for the receiver must overlap the fluorescence emission spectral peak of the donor. These conditions are met in the thylakoid, where the pigment is in high concentration, and the absorption spectra clearly overlap (see Figure 2.7). (Fluorescence is only seen from chlorophyll *a*.)

It is possible to measure the *action spectrum* for photosynthesis, meaning the number of molecules of product yielded per quantum absorbed plotted against the wavelength of the light. The action spectrum should fit the absorption spectrum of the pigment responsible for the photosynthetic process. Although these measurements are technically very difficult, and must be read with caution, it is clear that while the red end of the curve fits the *in vivo* absorption spectrum of chlorophyll *a*, the activity of photosynthesis follows other pigments in the region 450–600 nm, where chlorophyll absorbs relatively little light (see Figure 2.7). By this means it is concluded that at least some of the pigments listed above as accessory pigments do actually function in the manner suggested. This is true even for phycoerythrin in the red alga *Porphyridium*, where the spatial requirement is not obviously met as the pigment appears to be in granules, loosely attached to the thylakoids, as shown in Plate 8 (facing p. 84). Further evidence has been obtained by studying the fluorescence of chlorophyll *a* in organisms illuminated in regions where other pigments must absorb nearly all the light. Although chlorophyll in solution fluoresces at an intensity visible to the eye, while chlorophyll *in situ* does not, some fluorescent emission can be measured with the appropriate apparatus. Taking as an example *Porphyra*, light absorbed in the spectral range 520–560 nm, which is absorbed mainly by phycoerythyrin, promotes fluorescence in the region 680–690 nm, which must emanate from chlorophyll *a*. Moreover the intensity of fluorescence is as high or higher than that produced by direct illumination of chlorophyll *a*. Therefore in this case phycoerythrin is transferring energy to chlorophyll *a*.

The absorption spectra of the chlorophylls are sharp, and their contribution to the total absorption of the plant can be recognized easily. Other pigments have broader maxima, and identification of the components of the overall absorption or action spectra is much more difficult. The difficulty is increased by the fact that the quantities of pigments such as the xanthophylls diminish as the number of varieties increases. In the majority of species it is consequently not possible to say which of the pigments present are acting as accessory pigments. For the purposes of this text, it will be supposed that in general chlorophyll *b* where present has an accessory role, together with unspecified 'carotenoids' (either carotenes or xanthophylls or both).

There is no reason now to suggest that the primary pigment is other than chlorophyll *a* in green plants. However it appears that the arrangement of the chlorophyll molecules in the thylakoid, which is a very dense packing, leads to the formation of aggregates in which the absorption maximum is shifted slightly to the red, which is the reason why the absorption peak is much broader *in vivo* than when the pigments are extracted into solution. The shape of the *in vivo* spectrum is best explained by postulating special 'forms' of chlorophyll *a* absorbing at 670, 680, 690 and 700 nm (different laboratories disagree slightly in the wavelength allocations) which can be referred to as C_{670}, C_{680} and so on. The form C_{700} can reasonably be ascribed to photoreaction I, (see section 4.45) since II is hardly active at this wavelength. Again, the fluorescence at room temperature at 695 nm, which is associated with the system of photoreaction II, probably comes from C_{690}, which must itself therefore be ascribed to that reaction. The allocation of C_{670} and C_{680} is less easy. If these forms pass their energy to one or other of the forms C_{690}, C_{700}, then in a sense they are accessory pigments. This is leading to a concept of the photosynthetic pigment system as a funnel leading from a large quantity of a short-wavelength pigment through smaller quantities of longer wavelength pigments to reaction centres containing pigment absorbing furthest to the red and present in the smallest quantities. This funnel is the photosynthetic unit, the subject of the next section.

2.7 The photosynthetic unit

The concept of the *photosynthetic unit* is basic to present-day hypotheses about the mode of utilization of light energy. Each unit is supposed to contain a *reaction centre*, where the primary chemical process of photosynthesis is carried out one molecule at a time, connected to a relatively large number of pigment molecules, any one of which may absorb a photon and pass the energy to the reaction centre. The chemical reaction is much slower than the events in the pigment mass (see Figure 2.4) and this allowed Emerson and Arnold (1932) to separate the two processes. They exposed algal cells to flashes of neon-light; the flashes were of saturating intensity, meaning that no change was observed when the intensity was increased, and they were of brief duration (less than 10^{-5} s). The rate of carbon dioxide uptake by the cells was observed during the experiment, as the flashing-frequency was varied. It was found that the photosynthetic carbon dioxide uptake reached a maximum at a flashing-frequency of some 100 s^{-1}; from this they concluded that the 'dark reaction' of photosynthesis had a half-time of $\frac{1}{50}$ s. Secondly, as the flashing-frequency was lowered from that value, although the net rate of carbon dioxide uptake fell, the *yield calculated per flash* increased to a maximum value at frequencies of the order of 1 s^{-1}. They argued that at each flash, each reaction centre was activated, and reduced one molecule of carbon dioxide. By the time that the chemical reaction was completed, the energy of the flash had died away, so that each flash resulted in one and only one molecule being reduced per reaction centre. The long dark time was necessary to ensure that all the reaction centres had completed the reaction energized by one flash before the next arrived. At this frequency, the number of molecules reduced per flash was equal to the number of reaction centres. The sample of algal cells was then analysed, and the number of chlorophyll molecules determined. Hence the size of the photosynthetic unit was found. The value given by Emerson and Arnold was about 2500 molecules of chlorophyll for the reduction of one molecule of carbon dioxide, and the values were closely similar for several species.

Since that time, the concept has been somewhat modified. The discoveries that two different kinds of light-reaction centre were required to explain some observations (see Chapter 6) and that the primary process of each was an electron transfer reaction have led to the present concept that each reaction centre moves one electron per flash, that each electron passes in turn through two reaction centres, one of each kind, and that four electrons have to make this passage during the reduction of one molecule of carbon dioxide to carbohydrate (and the production of one molecule of oxygen from water). Hence eight separate 'photoacts' are involved, so that we should expect 2500/8, say 300 chlorophyll molecules to be associated with each reaction centre. Alternatively, the two reaction centres may draw energy from a common double-unit of 600 chlorophylls.

Further evidence for the existence of such a photosynthetic unit has been obtained from the electron-microscopic observations of subunits ('quantasomes') in thylakoid membranes, by Park, which appeared to be of the size to carry 300 chlorophyll molecules. Although the subunit structure has turned out to be much more complicated, so that the term 'quantasome' has lost its precision, there is a periodicity at this order of size.

The second line of work is based on measurement of the proportion of chlorophyll to other materials of the photosynthetic apparatus. Obviously for this to be significant there must be good reason to believe that the materials chosen are an essential part of the photosynthetic process (evidence for this will be discussed later). The materials cytochrome f and plastocyanin occur at approximate proportions of one molecule of each per 300 molecules of chlorophyll a. Cytochrome f and plastocyanin are believed to act as one-electron transfer agents, implying that 300 chlorophyll molecules cooperate in the movement of one electron.

2.8 Reaction centres

The reaction centres where a chemical change occurs using the excitation energy of chlorophyll represent one of the more difficult topics of study. Not only are the processes carried out very rapidly, but also the centres may depend on a precise

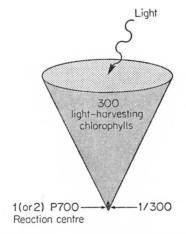

Figure 2.8. The light-harvesting and reaction centre chlorophyll of System I represented as a funnel

arrangement of several chlorophyll molecules, which may be easily deranged when attempts are made to isolate them. It appears that there are two photoreactions in photosynthesis (possibly more), and that their mechanisms differ. The centre for photoreaction I (defined in section 4.54) may operate by means of a specialized chlorophyll *a* molecule known as P700 (representing about 1/400 of the total chorophyll) which can be reversibly oxidized and reduced; one might regard the photosynthetic unit of photoreaction I as some 300 'light-harvesting' chlorophyll molecules with 1 or 2 'reaction-centre' chlorophylls (P700) (See Figure 2.8).

The centre for reaction II on the other hand is much more complex, as indicated by its much greater vulnerability to heat, disintegration of the thylakoid, and so on. It has been observed that this photoreaction is associated with the phenomenon of 'delayed light emission'. One attractive hypothesis is that the reaction centre consists of several chlorophyll molecules organized to form a semiconducting system, in which excited electrons could migrate freely, as could their 'holes' left at the ground level. On opposite sides of this structure would be electron and 'hole' traps, so that when the chlorophyll was illuminated electrons and holes would appear and fill the traps from which they would be taken by the electron transport chain of photosynthesis. If electron transport is prevented experimentally, the delayed light persists for seconds or minutes, and is stimulated by increasing the temperature. It is argued that the electrons and holes are dislodged from their traps by thermal energy back into the conduction band, in which they may combine with emission of light. This process is illustrated in Figure 8.2. In this concept there is light harvesting chlorophyll as before, but much more reaction centre chlorophyll. The traps may or may not be chlorophyll molecules.

References

Dougherty, R. C. and coworkers (1970). *J. Amer. Chem. Soc.*, **92**, 2826.

Emerson, R. and W. Arnold (1932). The photochemical reaction in photosynthesis. *J. Gen. Physiol.*, **16**, 191.

O'hEocha, C. (1966). Biliproteins. In T. W. Goodwin (Ed.), *Biochemistry of Chloroplasts*, Vol. 1, Academic Press, London, p. 407.

Hill, R. (1963). Chlorophyll. In M. Florkin and E. H. Stotz (Eds.), *Comprehensive Biochemistry*, Vol. 9, Elsevier, Amsterdam, Chapter 3, p. 73.

Strain, H. H. (1966). Fat-soluble chloroplast pigments. In T. W. Goodwin (Ed.), *Biochemistry of Chloroplasts*, Vol. 1, Academic Press, London, p. 387.

Kamen, M. D. (1963). *Primary Processes in Photosynthesis*, Academic Press, New York, p. 4.

Thomas, J. B. (1965). *Primary Photoprocesses in Biology*, North-Holland, Amsterdam.

Seliger, H. H. and W. D. McElroy (1965). *Light, Physical and Biological Action*, Academic Press, New York.

Smith, J. H. and A. Benitez (1955). *Modern Methods of Plant Analysis*, Vol. 4, Springer, Berlin, p. 142.

Tanada, T. (1951). *Amer. J. Bot.*, **38**, 276.

CHAPTER 3

Light energy into chemical energy

In the previous section a scheme was set out wherein light was absorbed by pigments, carotenes, xanthophylls and chlorophylls, producing a state of excitation which was transferred to chlorophyll *a*, or bacteriochlorophyll as appropriate. The excitation was stored at some site in the chlorophyll *a* mass, and this store was in contact with a photochemical reaction centre. In this section we will consider the way in which the energy of the excitation might appear in a chemical form.

3.1 Photochemistry

On pages 21–22 four types of chemical change were listed by which an atom or molecule could lose its excitation energy. Such reactions constitute photochemistry.

The fundamental 'laws of photochemistry' upon which the foregoing principles are based, have been variously stated, but the following may provide a convenient summary.

(1) The Grotthuss–Draper law, which states that only those radiations which are absorbed by the reacting system are effective in producing chemical change. This is not to be taken as implying that all or indeed any absorbed radiation is necessarily effective.

(2) Einstein's law of the Photochemical Equivalent, which states that each molecule of a substance taking part in a chemical reaction which is a direct result of the absorption of light takes up one quantum of the radiation causing the reaction. This 'direct' result may need to be applied in such a way as to exclude reactions subsequent to the formation of the excited state. This is a similar principle to that of E. Warburg, namely that a photochemical reaction depends on the number of absorbed quanta, not on their energy content. In more modern terms, remembering that the basis of photochemistry is the excitation of an electron by a photon, we may restate Einstein's law in the form: 'The basis of a photochemical reaction at the

molecular level is the activation of one electron by one photon'. This approach is well suited to the discussion of the photosynthetic unit in section 2.7.

A convenient quantity to measure in a photochemical reaction is the *quantum yield*, which is the number of molecules of a given substance formed following the absorption of one photon. (This quantity is often confused with its reciprocal, the *quantum requirement*.) The quantum yield gives an indication of the nature of the reaction. Thus very low yields imply that the chemical pathway accounts for only a small part of the total number of excited molecules, most of which are decaying (being *quenched*) by other processes. Very high quantum yields on the other hand are characteristic of chain reactions such as that between hydrogen and chlorine in which only the first step is activated by light.

3.11 Photochemistry of chlorophyll

Four kinds of photoreaction were listed in section 2.22. Of these, association and dissociation are not known for chlorophyll, neither is there any process known which might be compared to the photosensitization of H_2 dissociation by mercury atoms. All the photochemistry of chlorophyll appears to be derived from primary processes in which electrons are exchanged with other substances (oxidation–reduction reactions). Since this is in accord with what has been said so far about the primary process of photosynthesis, there has been considerable work done on the photochemistry of chlorophyll.

Chlorophyll in solution can be reversibly reduced or oxidized when illuminated in the presence of suitable electron donors or acceptors. A useful summary is given by Krasnovsky (1969), who discovered the reduction of chlorophyll to a pink compound of unknown chemical structure. The electron donor was ascorbic acid (vitamin C), and pyridine was used as the solvent, with the exclusion of oxygen. Krasnovsky and his group showed that the primary process was the formation of free-radical ions:

$$Chl^* + AH \rightarrow \cdot Chl^- + \cdot AH^+$$

that is, an electron has passed from the ascorbic acid (AH) to the chlorophyll (Chl) resulting in the formation of the positive and negative charges, and both molecules now possessed an unpaired electron, which is the essential definition of a free radical. Reference to Figure 2.5 shows that there is in chlorophyll a large system of conjugated bonds, that is, alternating single and double bonds, which fuse so that the orbitals of the electrons that form these bonds cover continuously all the carbon atoms of the ring system. This conjugated system probably holds the extra unpaired electron in the ion-radical $\cdot Chl^-$. This radical stabilizes itself, forming the pink product. Chlorophyll is easily regenerated using single-electron acceptors such as oxygen, quinones, riboflavin, methyl viologen and others. Oxygen, which possesses two unpaired electrons in the O_2 molecule, is a biradical itself; the other materials listed are able to accept single electrons, forming free radicals such as *semiquinone:*

Many reducing agents will act as electron donors to chlorophyll; in addition to ascorbic acid, cysteine, some ferrous compounds, hydroquinone and several others have been used, although in the case of hydroquinone a stable reduced product does not appear. The earlier photoproducts tend to lose their magnesium, forming phaeophytin.

Rabinowitch and Weiss in 1937 reported that solutions of chlorophyll in methanol or other solvents, illuminated in the presence of ferric chloride, became more-or-less reversibly bleached, indicating an oxidation of chlorophyll by ferric chloride. A similar photo-oxidation of the pigment has been found with oxygen, quinones, nitro compounds and methyl viologen, and as before the immediate photoproduct is an ion-radical, this time with a positive charge:

$$Chl^* + BH \rightarrow \cdot Chl^+ + \cdot BH^-$$

In the presence of oxygen the unstable intermediates form peroxides, which decay with emission of light (chemiluminescence) of the same wavelength as chlorophyll fluorescence, and the pigment molecule is degraded, at ring V or elsewhere. This process, and the phaeophytin formation referred to previously, are a warning to the experimenter to protect his preparations from air and light!

3.2 Primary photoproducts in photosynthesis

P700. There are two (or more) photoreactions in green plant photosynthesis, and probably in bacteria also. Reaction I (defined in section 4.54) appears to be much more like one of the photochemical systems than reaction II. As the electron acceptor is reduced, there appears in the chloroplast spectrum a diminution of absorbance at 700 nm, followed by its return. If a suspension of chloroplast fragments is carefully extracted with increasing concentrations of acetone, a stage is reached when a shoulder can be seen in the now depleted chlorophyll spectrum. This shoulder is reversibly removed by oxidizing agents such as ferricyanide, and restored by reducing agents. No pigment apart from normal chlorophyll (absorbing at 663 nm in acetone) is found when the fragments are completely extracted, and it is believed that the material absorbing at 700 nm is chlorophyll *a* modified by its environment; the bleaching is due to its oxidation. This pigment was named P700 by its discoverer, B. Kok. It is tempting to regard P700 in the oxidized state as being similar or even identical with the cation-radical formed during the photo-oxidation of chlorophyll with ferric chloride in methanolic solution. This is not proved however; P700 may turn out to be other than chlorophyll, and its oxidation may be mediated through a more immediate photo-oxidized product. A similar observation has been made with photosynthetic bacteria such as *Chromatium*, where the form of bacteriochlorophyll corresponding to P700 is known as P890. Parson, in Chance's laboratory, has shown that the oxidation of P890 precedes that of any other substance, so that oxidized P890 may fairly be regarded as a primary photoproduct. Although in the case of green plants no component has been shown to be oxidized faster than P700, there is some debate whether P700 is a necessary intermediate in the electron transport pathway.

Photoreaction II.   In the second photoreaction of photosynthesis, no such photoproduct as P700 can be observed. Witt (see, for example, Witt, 1969) has observed a spectral change at 682 nm, which he ascribes to the chlorophyll *a* belonging to photoreaction II, and which appears in less than 2×10^{-8} s. At the same time another peak appears with the same speed at 320 nm which may be identified with the formation of the semiquinone of plastoquinone. There is here the basis for a photo-oxidation of chlorophyll, plastoquinone being the electron acceptor. If this were, simply, the case, then it would follow that the centre of photoreaction II was a single chlorophyll molecule, in contradiction to the concept, based on the delayed light observations, of a larger formation. There is insufficient information even to make a guess, but the situation in reaction II is certainly more complex than in reaction I.

3.21 Application of photochemical studies to photosynthesis

The electron donors, which result in the photoreduction of chlorophyll, have very little effect on the fluorescence of the pigment, whereas the electron acceptors quench fluorescence effectively. Since fluorescence comes from the singlet excitation state, it appears that photoreduction proceeds exclusively via a triplet state, and photo-oxidation at least partly via the singlet. However nearly all photochemical reactions with dyes in solution proceed via the triplet state, for the reason that collision with a reactant molecule is unlikely during the short (10^{-9} s) lifetime of the singlet, but is possible during the much longer life of the triplet state. In fact, the kinetic analysis of both photoreactions indicates clearly that the triplet is involved in each case. However in photosynthesis the triplet state is unlikely to be involved for the following reasons. First, the triplet has not been directly observed. Secondly, the observed time for the earliest chemical change is of the order of 10^{-8}–10^{-9} s, which does not need the triplet lifetime for its explanation. Thirdly, the fluorescence of chlorophyll *in vivo* is some ten times less that *in vitro* (in solution) so that the singlet state *in vivo* is being deactivated early in its life, which would lessen the already small probability of the electron spin reversal taking place. Lastly, since triplet formation competes with other drains on the singlet state, triplet-activated reactions must have a lower quantum yield than those using singlet excited states. However, measurements of the quantum yield of photosynthesis approaching $\frac{1}{8}$ (molecule of CO_2), by several workers, is believed to represent almost 100% efficiency of energy conversion, and under these conditions there can be little competition from the other energy drains.

However, although the triplet state may be disregarded as an intermediate in photosynthesis, the photochemical systems have important applications. Thus in photoreaction I of photosynthesis, ferredoxin, an electron acceptor, is reduced, and P700, which appears to be a form of chlorophyll *a* is oxidized. The photo-oxidation systems may provide a model for this process. Secondly, in one theory of the reaction centre of photoreaction II it is regarded as containing an aggregation of chlorophyll molecules, which separates charge by a photoconduction process (see Figure 8.2). If the charge appears on chlorophyll molecules (the charge traps) then these molecules would be seen by other molecules as anion and cation-radicals, just as in the photochemical systems. Furthermore, since these charge-traps have to initiate a chain of electron transport reactions (see Chapter 4) it is of great importance to know the *redox potentials* (see section 4.2) of the traps in relation to the rest of the chain, and the photochemical systems do allow an approach to be made to the measurement of the chlorophyll derivatives involved. A discussion of the estimated values of the redox potentials is given in Chapter 4.

3.3 Light energy and separation of charge

The absorption spectrum of chlorophyll in Figure 2.2 was explained in terms of the formation of singlet excited states, the bands in the red and blue forming the first and second excited states respectively. The second excited state is converted in a radiationless transition to the first. The position of the absorption maximum in the red gives an indication of the mean energy involved in the promotion of an electron from one of the vibrational levels of the ground state to one of the first excited state. The figure is obtained using the equations (see section 1.2)

$$E = h\nu = hc/\lambda$$

Taking h (Planck's constant) as $6 \cdot 626 \times 10^{-27}$ erg s and c (the speed of light) as $2 \cdot 998 \times 10^{10}$ cm s^{-1} we can attach energy values to the wavelength scale of Figure 2.2. The above equations give the energy in ergs per photon (or per excited molecule). A more directly useful unit than the erg is the electron volt (the work done on an electron passing through a potential rise of one volt), equal to $1 \cdot 602 \times 10^{-12}$ erg. Since we shall be concerned in the next chapter with the redox potentials of substances taking part in photosynthetic electron transport, it is convenient to have potential measurement in volts for the energy of the excited electron of chlorophyll. On the other hand, the process of photosynthesis is directed at storing the energy of light in chemical materials, and for this purpose we find the calorie* per mole a more informative scale. 1 eV is equivalent to $23 \cdot 06$ kcal mole^{-1}. Figure 3.1 sets out a suitably annotated version of Figure 1.2.

Chlorophyll *in vivo* is organized, however, into various aggregates which have absorption maxima displaced toward the red; we can assign a value for the energy of the singlet excited state in each case. The overall process of photosynthesis diminishes in efficiency at wavelengths of light above 680 nm and is virtually extinct at 700 nm; however this appears to be due to failure only of photoreaction II, photoreaction I being still detectable at wavelengths up to 720 nm or longer. It has been suggested that the 700 nm limit on photoreaction II is the practical limit of the 'tail' of the absorption of the chlorophyll a form C_{690}, and the 720 nm limit on reaction I likewise indicating the extent of the absorption of the form C_{700} (which may include P700). It should be noted that 720 nm illumination, which has an energy of $1 \cdot 72$ eV, can form an excited state of chlorophyll C_{700}, with energy $1 \cdot 76$ eV. The extra $0 \cdot 04$ eV is provided by the thermal energy of the pigment; the longer wavelengths only excite the 'hotter' molecules.

If the light energy that operates photoreaction I is indeed channelled through the pigment absorbing at 700 nm, then the energy available at the reaction centre is $1 \cdot 76$ eV per photon, and likewise $1 \cdot 80$ eV per photon for reaction II operating through C_{690}. This is the energy available for each photoreaction respectively. The excited electron, while still electrostatically under the influence of the positive charge of the $\cdot$Chl$^+$ (if that ion-radical is indeed

* The joule rather than the calorie is now preferred as the unit of energy but it is felt that no service will be done to the student by introducing it here.

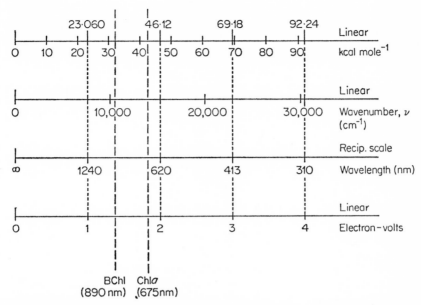

Figure 3.1. The equivalence of four measures of energy commonly used in photochemistry. From Clayton, R. K. (1965) *Molecular Physics in Photosynthesis*, Blaisdell, New York, Fig. 3.1, p. 39, with permission

formed) can move away some distance into the mass of surrounding material, and find a resting place in the form of an electron acceptor (see Figure 3.2).

The primary act of photosynthesis therefore established *a charge separation* across a certain distance. The same result follows from the semiconductor

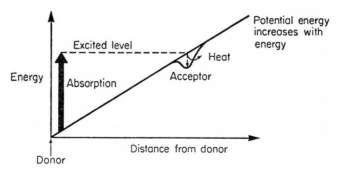

Figure 3.2. To illustrate the trapping of an electron from an excited state at some distance from its origin.

model of the reaction centre II (Figure 8.2): the two traps, for electrons and holes, must be separated by a certain distance for the mechanism to work. This separation of charge means that an electrical field will be set up. Suppose that the potential difference is of the order of 1 V, and the distance is of the order of the thickness of the thylakoid membrane, say 10^{-6} cm, then the field is some 10^6 V cm^{-1}, which is very large by microscopic standards. One possible consequence of this is that the absorption spectra of the thylakoid membrane components will be altered by the electrical stress (the Stark effect). Witt's group in Berlin have indeed observed a small change in the spectra of chloroplasts (known as the 515 nm shift) which appears fast: in the region of 10^{-8}–10^{-9} s. They regard this as a Stark effect of an internal field on the chlorophylls and carotenoid pigments.

3.4 Summary

Light quanta are absorbed by chlorophyll and passed to reaction centres, where charge is separated either by loss of an electron to an acceptor (or gain from a donor), or by the more or less simultaneous appearance of opposite charges on different chlorophyll molecules separated by some distance. The former and possibly the latter concept may be clarified by work on positively and negatively charged forms of chlorophyll formed as intermediates in photoreactions in solution. P700 may be a positively charged chlorophyll ion-radical of that type. The energy taken to form the excited state of chlorophyll *a in vivo* is some 1·8 eV, or 42 kcal mole^{-1}. A spectroscopic change at 515 nm may be due to the strong electric field brought about by the charge separation.

References

Clayton, R. K. (1965). *Molecular Physics in Photosynthesis*, Blaisdell, New York, p. 39.

Krasnovsky, A. A. (1969). The principles of light energy conversion in photosynthesis: Photochemistry of chlorophyll and the state of pigments in organisms. In H. Metzner (Ed.), *Progress in Photosynthesis Research*, Vol. 2, Institut für Chemische Pflanzenphysiologie, Tubingen, p. 709.

Mahler, E. H. and H. R. Cordes (1966). *Biological Chemistry*, Harper and Row, New York, p. 208.

Rabinowitch, E. and J. Weiss (1937). Reversible oxidation of chlorophyll. *Proc. Roy. Soc. A*, **162**, 251.

Witt, H. T. and coworkers (1969). Evidence for the coupling of electron transfer, field changes, proton translocation and phosphorylation in chloroplasts. In H. Metzner (Ed.), *Progress in Photosynthesis Research*, Vol. 3, Institut für Chemische Pflanzenphysiologie, Tubingen, p. 1361.

CHAPTER 4

Electron transport

According to the model used for this introduction to photosynthesis, light energy is absorbed by pigments, and the excitation is channelled via chlorophyll *a* (or bacteriochlorophyll) to reaction centres. In green plants there are two centres, in bacteria possibly only one. At the reaction centre the energy of the excited pigment causes an electron to be transferred from a donor (which becomes oxidized) to an acceptor (which becomes reduced). Some of these primary photoreactants have been tentatively identified. In this section the principles of electron transport will be discussed, and the view will be set out in which the solid part of the photosynthetic apparatus has the role of providing an electron transport pathway in which the two light-reactions of green plant photosynthesis function in series. This view is adopted because it gives some account of most observations that have been made, and because it is held as a working hypothesis by a large majority of biochemists. There are alternative hypotheses, most of which offer more satisfactory interpretations of parts of the data, and these will be discussed in a later chapter.

4.1 Redox couples

Two compounds which are interconvertible by means of an oxidation–reduction reaction are said to be the two forms of a 'redox' substance or 'redox couple'. Although the original use of the term oxidation meant a combination with oxygen as in the formation of an oxide, most oxidations in biochemistry can be expressed as a transfer of hydrogen atoms or electrons:

$$AH_2 + B \rightarrow A + BH_2 \qquad (4.A)$$

$$X \rightarrow X^+ + e^- \qquad (4.B)$$

In reaction (4.A), AH_2 reduces B by transfer of hydrogen atoms. In reaction (4.B), half a reaction is shown in which X acts as a reductant or electron donor. A second half-reaction would of course be required since free or solvated electrons are not known to occur in biological systems. The oxidation of succinate $(AH)_2$ to fumarate (A) with the concomitant reduction of the flavoprotein enzyme succinate dehydrogenase $(B \rightarrow BH_2)$ follows reaction (4.A), and the oxidation of cytochrome c from the ferrous to the ferric form follows reaction (4.B).

4.2 Standard oxidation–reduction potentials

Redox couples can be arranged in a linear series. Their relative positions are described in terms of a scalar quantity known as the *standard oxidation–reduction potential*. From these potentials one can describe and predict the

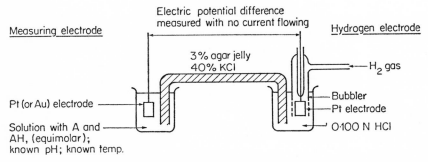

Figure 4.1. The potentiometric determination of redox potentials

changes when two couples are allowed to react. The standard potential is that which would be measured between the connections to an electromotive cell made up of one half-cell containing equimolar proportions of the oxidized and reduced forms of one couple, and a standard half-cell. Various types of standard half-cells are in use, but the ultimate standard of reference is the 'normal hydrogen electrode'. This is a platinum surface in contact both with an acid solution of unit activity (approximately 1N) and hydrogen gas at 1 atmosphere pressure. The reference and the 'unknown' half-cells are connected by a 'salt bridge' of saturated potassium chloride as shown in Figure 4.1. The pH of the first half-cell must be stated; pH 0 is preferred, but is not usually possible with biological material. The temperature must be defined; 25° is the preferred value. The potential is measured either by opposing it with a known potential such that no current flows, or by means of a device such as a pH meter which has an input resistance of the order of 10^{12} ohms. The sign of the potential is that of the lead from the experimental half-cell. This is the 'European' convention. The symbol is E_0 or E^0. If the pH is other than zero, the symbol E'_0 is used, with an indication of the actual pH value. At pH 7 the standard potential E_0' of the hydrogen electrode is diminished by 7×60 mV to -0.420 V at $25°$ (see Equation (4.3)).

If the ratios of the oxidized and reduced forms of the couple are varied, the potential E is no longer the standard value (E_0), and is given by

$$E = E_0 + \frac{RT}{n\mathrm{F}} \ln \frac{\text{[oxidized form]}}{\text{[reduced form]}} \qquad (4.1)$$

In this equation n is the number of equivalents per mole, for example two in reaction (4.A) and one in (4.B). F is the Faraday, equal to 96 500 coulombs per gram-equivalent, R is the gas constant, 8.314 joules deg^{-1} mole^{-1}, and T is the absolute temperature. Then at $25°C$ and using decadic logarithms

$$E = E_0 + \frac{0.05915}{n} . \log \frac{\text{[oxidized form]}}{\text{[reduced form]}} \qquad (4.2)$$

The numerical factor increases by 0.0002 per degree over the physiological range. For most purposes 0.05915 V can be written 60 mV or 0.06 V.

If a reductant of the AH_2 type can dissociate as an acid:

$$AH_2 \rightarrow AH^- + H^+$$

$$AH^- \rightarrow A^{2-} + H^+ \qquad (4.C)$$

then the redox potential will vary with the pH. This dependence is numerically similar to Equation (4.2)

$$E_0' = E_0 + a . \frac{0.05915}{n} . \log [H^+]$$

$$= E_0 - a . \frac{0.05915}{n} \quad \text{(pH at 25°C)} \qquad (4.3)$$

The term a/n in Equation (4.3) is the number of hydrogen ions released per electron transferred, so that with a dibasic acid of the type AH_2, a/n has three values: zero when the pH is considerably above the second pK, $\frac{1}{2}$ for pH values between the two pK values (provided these are separated by at least 2–3 pH units), and unity for pH values below the first pK. Within about one pH unit on either side of each pK value, a/n has a varying fractional value so that a smooth curve results for E_0' plotted against pH. A typical curve is sketched in Figure 4.2.

The above treatment of standard potentials needs a further qualification with materials such as cysteine, which in the oxidized form (cystine) is dimerized. This means that not only does the term 'equimolar proportions of oxidized and reduced forms' cease to mean the same as '50% reduction', necessitating an *ad hoc* specification of the 'standard' cysteine–cystine

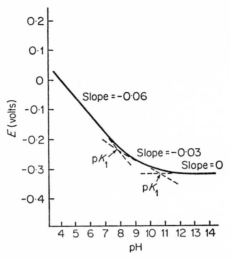

Figure 4.2. The dependence of E_0' on pH for anthraquinone-2,7-disulphonate. From E. H. Mahler and H. R. Cordes (1966), *Biological Chemistry*, Harper and Row, New York, Fig. 5.6, p. 208, with permission

couple, but also that a squared term appears in Equation (4.1)

$$E = E_0 + \frac{0 \cdot 06}{n} \log \frac{[\text{cystine}]}{[\text{cysteine}]^2}$$

4.21 Variation in the proportion of oxidized to reduced species

From Equation (4.1), a tenfold change in the proportions of oxidized to reduced species causes a change of $0 \cdot 06/n$ in the potential. This sets practical limits on the oxidizing power of a redox material in the same way as there are limits to the buffering range of a weak acid in the presence of its salt. Just as it is convenient to formulate a pH buffer within the range of 1 pH unit on either side of the pK of the weak acid, so that the ratio of acid to salt is within the range 0·1–10, so the corresponding range for a redox buffer is $(0 \cdot 06/n)$ V. With present day methods of analysis, a material can be said to be 'completely' oxidized or reduced when the ratio exceeds 10^2 or 10^3 to one. If therefore a gap of 0·24 V separates the standard potentials of two redox materials, the lower can have no significant oxidizing action upon the upper if n is unity for both; if n has a higher value, the necessary gap is correspondingly less. For practical purposes, a couple which has a higher redox potential will oxidize a couple with a lower one. Where the difference is less than $0 \cdot 24 \, N/n$, a measurable equilibrium will be set up—provided any necessary catalyst is present.

Provided that conditions of pH, temperature and concentration are considered as and where appropriate, the standard oxidation–reduction potential of a redox material, and the relative standard potentials of a group of redox materials, are of the greatest value in elucidating the biochemical role which such groups play.

4.3 Coupled oxidations

If two separate redox couples are mixed and they react, they must reach a common potential by adjustment of the ratios of the oxidized to reduced species of each couple, each according to Equation (4.1). If the two couples do not react, they may yet be brought to a common equilibrium potential by the addition of a third redox material with which they both react. In this case the oxidation or reduction of one material by a second is said to be coupled by the third.

For substances in free solution, it is rare to find systems of more than four such redox systems coupled together. In such a system, the standard potentials must either increase from left to right, or else, if there is a step in the reverse direction, the considerations of the previous section must limit it to the order of 0·1–0·2 V. One example is the reaction

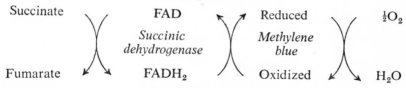

In the above example the standard potentials are (at pH 7): succinate–fumarate, 0·03 V, FAD–FADH$_2$ (in succinic dehydrogenase), approximately 0 V; methylene blue (oxidized–reduced), 0·01 V, and oxygen–water, + 0·82 V.

For the sake of brevity, redox couples will be referred to from now on more briefly; the potential of cytochrome f will stand for the potential of the oxidized–reduced cytochrome f couple. There is seldom more than one redox reaction associated with each substance that we shall be considering.

In the above example electron transfer took place from a couple of lower standard potential (succinate–fumarate) to one of higher potential (oxygen–water). This is the natural or 'thermodynamic' direction (see Figure 4.3). Electron transfer in the reverse direction requires a supply of energy, either to be supplied by adjustment of the concentration ratios of the reactants so as to bring their actual potentials into the proper order, or by introducing a simultaneous additional chemical change. Chemical changes that are 'coupled' to an oxidation–reduction reaction are common in enzyme-catalysed biochemical processes, and often involve the formation or breakdown of nucleoside triphosphates:

$$AH_2 + B + ADP + P_i \rightarrow A + BH_2 + ATP \qquad (4.D)*$$

The coupled reaction has a characteristic *standard free energy change* ΔF_0, the actual free energy change, ΔF, depending on the concentrations of the reactants and products, and on the pH. This free energy change superimposes

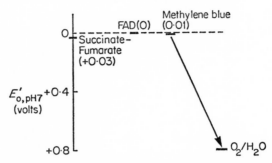

Figure 4.3. Diagrammatic representation of an electron transport process, using a scale of standard redox potentials.

on the potential difference, E, for the oxidation–reduction reaction, a potential ΔE related to the free energy change by the expression

$$\Delta F = -nF \, \Delta E \quad \text{where F is the faraday,}$$

96 649 joules mole^{-1} (23 088 cal mole^{-1}). In example (4.D), taking a free energy change of -8000 cal mole^{-1} for the formation of ATP under physiological conditions, the change in the redox potential difference for the couples AH_2–A and B–BH_2 is 0·173 V (in this example $n = 2$).

4.4 Electron transport in biological systems

In living cells, there are solid systems in which oxidation–reduction reactions are believed to occur by means of several steps in series, each involving a specific redox couple. Examples are the inner membranes of mitochondria, the membranes of the endoplasmic reticulum obtained as 'microsomes', the outer and inner membranes of the nuclear envelope in certain cells, chloroplast thylakoids in green plants, thylakoids in the blue–green algae, and

---

* ADP represents adenosine diphosphate, P_i represents the orthophosphate ion, ATP represents adenosine triphosphate. The water formed in the condensation is not usually shown in the equation; there is also an uptake of hydrogen ions, which is not shown either.

membrane preparations from bacteria which may have both oxidative and photosynthetic roles. Probably the system which has been studied at the greatest intensity is the mitochondrial pathway of oxidative phosphorylation, in which reduced nicotinamide adenine dinucleotide (NADH) and succinate are oxidized, oxygen is reduced to water, and ADP is phosphorylated to ATP in fresh preparations. For details of the mitochondrial 'respiratory chain' reference may be made to Klingenberg (1968). Figure 4.4 reproduces a diagram summarizing the process, in the same form as Figure 4.3. The

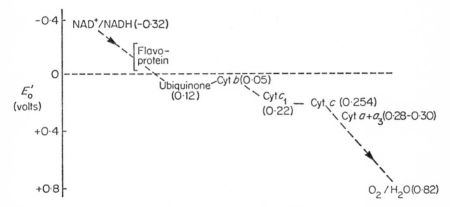

Figure 4.4. Representation, in the same way as in Fig. 4.3 of the electron transport in the mitochondrial respiratory chain. By permission of M. Klingenberg

reactions of the chloroplast will later be drawn on such a diagram for comparison (Figures 4.7 to 4.10).

The inner membrane of the nuclei of certain cells appears to possess a respiratory chain of the mitochondrial type; the outer membrane and the microsomal system possess pathways of electron transport which are less well understood. For a discussion of these the reader is referred to Strittmatter (1968). It is not established whether the cell membrane (cell envelope) has electron transport pathways, but the same membrane in bacteria, which appears to be the only membrane system present, is believed to carry out the functions of the mitochondria of eucaryotic cells. Smith (1968) gives a useful review of this topic.

It happens that the redox components of these membrane systems—transition-metal compounds, quinones, flavoproteins and so on—have different absorption spectra according to whether they are in their oxidized or reduced states. Provided that the biological material is sufficiently transparent to light (visible or ultraviolet) in the region of the difference, the

3

changes in the redox state of a given component can be observed directly by a spectrophotometer during the operation of the electron transport pathway. By observing two or more components simultaneously in this way, the order in which they react can be established. Reference has already been made to this in section 3.2.

Often the first indication of the presence of a redox component in a biological structure is the observation of a change in the absorption spectrum during an oxidation–reduction reaction. The membrane material is then fractionated and the unknown material purified and characterized. Redox materials seem to fall into relatively few families, and the chemical nature of many redox intermediates can be inferred directly from the difference spectrum of the oxidized and reduced states of the original material.

After the discovery of the electron transport system in the mitochondrion, other sites were discovered as described above. The concept of electron transport includes three features: first, there must be several, usually at least three, redox components present; second, most if not all of the intermediary components must be bound together in one 'solid' matrix such as a membrane; and third, it must be shown that electron transfer takes place in a sequence from one to the next.

These features are found in photosynthetic systems.

4.5 Electron transport in photosynthesis

Fresh preparations of chloroplasts from peas and other plants are capable of taking up carbon dioxide when illuminated (whether as the dissolved gas or as bicarbonate ions will be discussed later) and forming carbohydrate; oxygen is evolved in this process. This provides a reason, in this text, for concentrating biochemical attention on the chloroplast as the site of photosynthesis. For a treatment of the problems involved in the maintenance by the organism of the necessary conditions for the chloroplast to do this *in vivo* reference may be made to Rabinowitch (1945, 1951). We will be concerned with the control that the cell undoubtedly exerts over the photosynthetic process in the chloroplast. However for the present it is sufficient to state that the chloroplast carries out the process described by the equation, first derived by de Saussure:

$$CO_2 + H_2O \rightarrow (CH_2O)_n + O_2 \tag{4.E}$$

This process is clearly the reverse of the overall equation describing the respiration of carbohydrate:

$$(CH_2O)_n + O_2 \rightarrow CO_2 + H_2O$$

Equation (4.E) is an oxidation–reduction reaction, and all the component reactions which it summarizes are located within the envelope of the chloro-plast in plant cells.

It is widely accepted that the metabolism of carbon, from carbon dioxide to carbohydrate, takes place in the solution (stroma) phase of the chloroplast. The evidence is not complete, and alternative points of view will be discussed in a later chapter. For the present purpose, however, the above hypothesis will serve. The carbon metabolism requires a supply of reduced NADP and ATP besides carbon dioxide, enzymes, cofactors and catalytic substrates. There is some analogy here with the oxidative process of the mitochondrion, which takes place mostly in the stroma (matrix), and the NADH is oxidized by the cristae, which produce ATP. It is established that given only one soluble inter-mediate, the protein ferredoxin, the thylakoids of the chloroplast reduce NADP and produce oxygen and ATP. In the analogy, the cristae of the mitochondrion correspond with the chloroplast thylakoids.

For the rest of this chapter we shall examine the redox materials of the chloroplast, their properties and location. A scheme can be drawn showing possible reaction pathways, but the evidence for it will be discussed later.

4.51 The redox components of photosynthetic systems

Small molecules. Nicotinamide coenzymes. NADP* (Figure 4.5,I) is the chloroplast coenzyme. NAD is not known to occur in chloroplasts of green

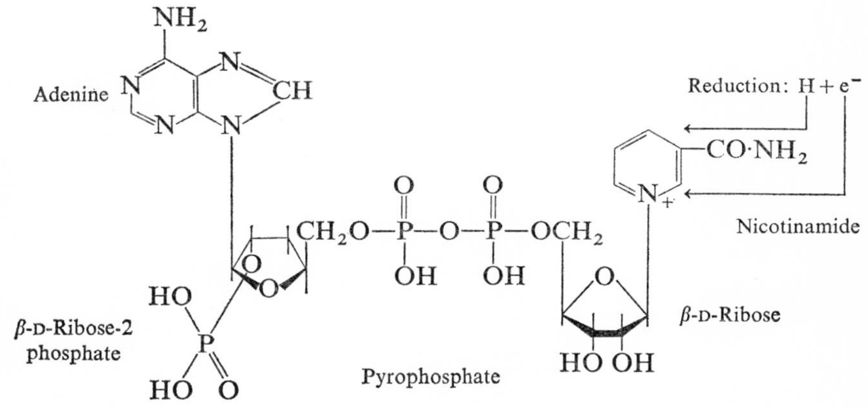

I NADP

Figure 4.5. Formulae of some redox materials important in the study of photo-synthesis

---

* The convention to be followed here is that the acronym NAD or NADP does not indicate the redox state: the forms NAD^+, NADH, $NADP^+$, NADPH indicate the oxidized and reduced states respectively.

Figure 4.5 (*contd.*)

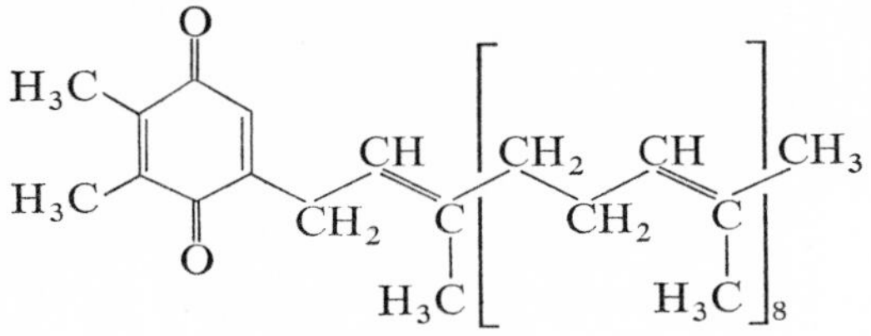

II Plastoquinone (Kofler's quinone; Q 254; PQA-45)

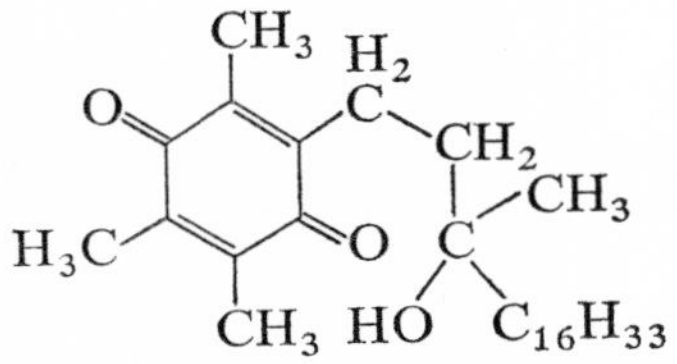

III α-Tocophenylquinone

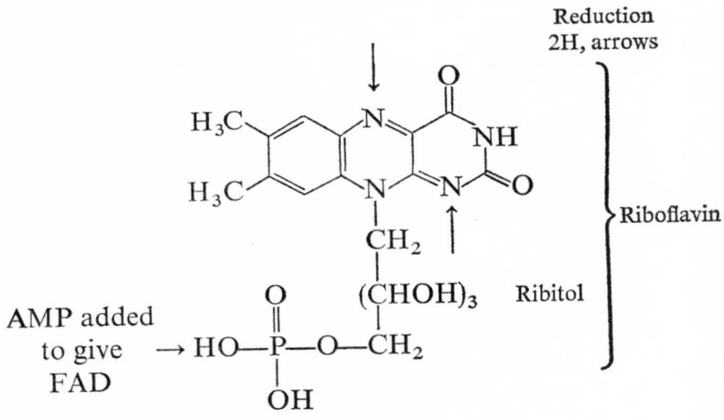

IV FMN

plants. In the blue green algae, where there is no distinction between 'stroma' and cytoplasm, the roles of NAD and NADP in photosynthesis are uncertain. In bacteria it is probable that NAD is important.

Quinones. The principal quinones are the plastoquinones (Figure 4.5.II) and tocophenylquinones (Figure 4.5.III). They are only soluble in hydrocarbons or absolute ethanol, where their redox potentials, measured indirectly, may have only an approximate relevance to their environment *in vivo*. *Lipoic acid* is present, but is seldom assigned a role. *Manganese* ions, or an easily dissociated complex of manganese, can be detected and has an ascribed role.

Flavins. Flavin mononucleotide (Figure 4.5.IV) is used as a reagent but does not appear to occur free, nor free as the adenine dinucleotide form. A number of flavoproteins have been isolated from plants.

Protein conjugates. Conjugated proteins play a large part in electron transport, and biological oxidations in general. The oxidoreductase enzymes activate a specific metabolic substrate and transfer electrons or hydrogen atoms between it and another metabolite or coenzyme. The coenzyme may be sufficiently tightly bound to the protein that it can be regarded as a prosthetic group. Not all these proteins are enzymes; if such a conjugated protein can accept (or donate) electrons from a wide variety of sources, then activation of specific substrates is no longer part of the mechanism, which is what many biochemists would consider essential to the definition of an enzyme. It is with this group that this section is concerned.

The redox properties of these conjugated proteins lie in the nature of the prosthetic groups, and the proteins can be grouped in this way. The prosthetic group, when separated from the protein, has a characteristic redox potential, but *in situ* this is raised or lowered by the influence of the protein. Thus the flavin nucleotides FMN and FAD have a standard redox potential E_0' (pH 7) at $25°$ of -0.185 V, whereas the *flavoproteins* can have values between zero (succinate dehydrogenase) and -0.3 V (ferredoxin–NADP oxidoreductase). While many flavoproteins can be isolated from leaves of plants, it is doubtful whether so many are concerned with photosynthesis or are even chloroplast components. The yellow enzyme ferredoxin–NADP oxidoreductase (NADP reductase) mentioned above is the only flavoprotein with a role in the scheme of photosynthesis being presented here. It is located in the thylakoid membrane, but is slowly leached out *in vitro*. The molecular weight is 40 000; there is one mole mole^{-1} of FAD and E_0' (pH 7) is less negative than NADP (-0.32 V).

Copper atoms or ions are often found in conjugation with proteins, such as haemocyanin in invertebrates, or the polyphenol oxidases. Usually there is an intense blue colour, indicating that $d–d$ transitions which were 'forbidden' in the simple ions, have become 'allowed' in the complexed state. The standard potentials are closer to that of the cupric–cuprous couple, the scatter being some 0.35 ± 0.05 V. In the chloroplast the protein *plastocyanin* (molecular weight 21 000, two atoms of copper) has an intense blue colour, a potential of $+0.370$ V, and apparently an important electron transport role in the thylakoid membrane.

'*Non-haem*' *iron* has recently been found to be a constituent of several enzyme systems, and is characterized by absorption in the blue end of the spectrum, release of ferrous iron by acid, and usually release of hydrogen sulphide at the same time. The nature of the prosthetic group is not certain, although a bridged complex of iron ions and sulphur atoms seems likely. The observed range of NHI proteins covers a potential span of from zero (in the succinate dehydrogenase complex) to the very low potentials of the *ferredoxins* of bacteria and plants. Plant (spinach, *Chlorella* and parsley) ferredoxins are located within the chloroplast, and have potentials of the order of -0.430 V at pH 7 and 25°. They are red to red–brown in colour and have molecular weights of 12–13 000, with two atoms each of sulphur and iron. Either ferredoxin is only very loosely bound to the thylakoid, or else it is a stroma protein; until recently the preparation of chloroplasts always involved some damage to the envelope, and ferredoxin was lost.

P700 is presumed to be a protein conjugate, on somewhat inadequate evidence. The measured potential is $+0.430$ V. There seems little point in comparing it with the potential of 0.645 V observed for chlorophyll oxidation by ferric chloride (in ethanol) until more is known of the nature of the reaction.

Haem conjugates. Of the three groups into which haem-proteins are usually divided, the haemoglobin group is not represented in the chloroplast. Of the catalase–peroxidase group, catalase is certainly represented, in both stroma and thylakoid, although the quantity is much less than in the cell cytoplasm. Nevertheless there is sufficient to cause a fast evolution of oxygen in the presence of 1 mM hydrogen peroxide. Catalase is only known in the ferri-haem state, although it can be reduced after partial denaturation. Although it is the only enzyme known which can evolve oxygen, and although it may possibly form redox couples of very low or very high potential, no direct role has been found for this enzyme in the electron transport system. Both catalase and peroxidase are found in bacterial chromatophores.

The remaining class of haem-proteins comprises the *cytochromes*. Two types are represented in thylakoids, cytochromes *b* and *c*. None of the *b* cytochromes have been isolated from thylakoids, and their identification has been by means of spectroscopic observation of the material, usually after removal of the chlorophyll. Cytochrome b_6 has a potential of -0.06 V, and a characteristic alpha-band in its spectrum at 563 nm. There is some reason to believe that it may have two components, one of which may absorb at 559 nm. Cytochrome b_3, with a potential of between 0.058 and 0.260 V, was originally isolated from plant microsomes, but it has been claimed that it exists in chloroplasts as well; it absorbs at 559 nm. A third cytochrome absorbing at

559 nm has recently been reported, which has a potential of 0·37 V, which is very high if it is a *b* cytochrome. However 559 nm is on the classical border between *b* and *c* cytochromes. This distinction is now made chemically according to the nature of the haem, since *b* cytochromes have acid-released protohaem-IX (see Figures 4.6.I and II) whereas *c* cytochromes have an acid-fast haem, covalently bound to the protein. Of the above, only cytochrome b_6 and microsomal b_3 are fully characterized as *b* cytochromes.

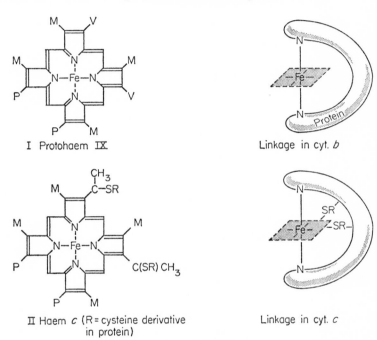

I Protohaem IX

Linkage in cyt. *b*

II Haem *c* (R = cysteine derivative in protein)

Linkage in cyt. *c*

Figure 4.6. The binding of haem in cytochromes of types *b* and *c*.

The *c* cytochromes are represented in chloroplasts in higher plants by cytochrome *f* (*f* standing for *frons*, a leaf), which has been isolated; it has a molecular weight of 120 000, two haem groups, and a potential of +0·370 V. The alpha-band is at 555 nm. In the green algae a very similar cytochrome is known as C552.

Preparations of bacterial chromatophores contain cytochrome c_2, which has a potential of +0·33 V, or other *c*-types. The range of molecular weights and potentials is great. The occurence of the *b*-type cytochromes appears to be limited to the Athiorhodaceae, and may not be photosynthetic. A 'hybrid', '*Rhodospirillum* haem protein, RHP' has been isolated from both *R. rubrum* and *Chromatium*. It has intermediate cytochrome properties, and a potential of approximately zero. The purple bacteria also contain cytochromes, C422 and C423·5, which have been mentioned previously; they are believed to lie close to the reaction centre.

4.52 Cyclic electron transport

In section 3.2 the qualified assertion was made that the result of the absorption of light by the chloroplast resulted in the formation of oxidizing and reducing agents as the first detectable step. If these entities were allowed to react with each other, either directly or by the mediation of a number of

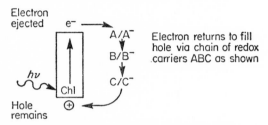

Figure 4.7. To illustrate the principle of cyclic electron flow, using the style of Fig. 4.3. (Compare with Figs. 4.8 and 4.9)

redox materials, a steady state could be visualized in which electrons were driven round a chemical circuit by the energy of light. One might detect a change in the redox balance of a given component when the light was switched on or off, but in the steady state there would be no release of oxidizing or reducing equivalents to the surrounding space (see Figure 4.7).

The value of such a cyclic process lies in the observation that passage of electrons through biological chains of redox substances is often associated with the formation of the energy-storage compound ATP. A typical case is that of the mitochondrion where the electrons originate from the oxidation of respiratory substrates such as succinate, and leave to form water from atmospheric oxygen. Hypotheses concerning the mechanism of the coupling of electron transport to phosphorylation of ADP are many, and their testing is a problem of general biochemistry. For the present, let it be assumed that whatever process holds in one case holds for all.

4.53 Non-cyclic electron transport

The life of the green plant depends on the reduction of carbon dioxide to carbohydrate by means of light energy. It is clear that in this case a reducing agent is being released by the photosynthetic process. At the same time oxygen is evolved, so that it is clear that non-cyclic electron transport takes place.

Unless the separated oxidizing charge resulting from the illumination of chlorophyll passes directly to water, either one or a chain of redox substances must be involved to couple the electron transport. This hypothetical chain is usually regarded as passive, and hence the standard potentials of the components must be around or above the potential of the hydroxyl-ion–oxygen reaction (+815 mV at pH 7). The light-induced oxidant, whose position is labelled Z in many diagrams, is hence more positive in potential than 810 mV (see Figure 4.8).

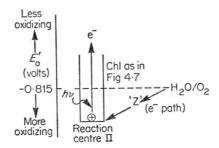

Figure 4.8. The oxygen-evolving reaction of photosynthesis in green plants

The Hill Reaction

There must be a light-induced reductant corresponding to the oxidant above. To complete a path for non-cyclic electron flow an electron acceptor must be available. Before the discovery of the natural acceptor, ferredoxin, other oxidizing agents were known, such as complexes of ferric iron (the uncomplexed ion Fe^{3+} is not soluble at physiological pH values) benzoquinone and indophenol dyes. In the presence of one of these oxidants, chloroplasts or chloroplast fragments will evolve oxygen when illuminated. This is the *Hill Reaction*, and the oxidants are *Hill oxidants*. It is subject to the 'red drop', that is, a decline in activity at wavelengths above 680 nm virtually reaching zero by 700 nm.

There is apparently another passive electron transport chain connecting the primary, light-induced reductant with the site at which the Hill oxidants act. Plastoquinones are likely to be involved, with cytochrome b_6 a less likely partner. Between that point and the primary reductant a site can be labelled X_{II} to indicate a possible unknown intermediate.

The Hill reaction proceeds at a greater rate if the chloroplasts are supplied with ADP, phosphate and magnesium ions. ATP is then formed. This indicates that phosphorylation is coupled to the electron transport of the Hill reaction, as opposed to the recombination of oxygen and reductant by some 'mitochondrial' system. Ammonium salts, and amines, release the coupling analogously to uncoupling agents like dinitrophenol in the mitochondrion. These uncoupling agents inhibit the Hill reaction in the chloroplast at higher concentrations. Phosphorylation of ADP accompanied by the oxidation and reduction of substrates is termed '*non-cyclic* photophosphorylation'.

There is evidence that manganese and chloride ions are required for the Hill reaction; this will be discussed later.

The reduction of NADP. The diphosphoglyceric acid reductase (3-phospho-glyceraldehyde dehydrogenase, EC.1.2.1.13) in the chloroplast stroma is

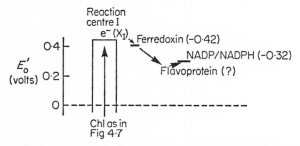

Figure 4.9. The NADP$^+$-reducing system; compare with Figs. 4.7 and 4.8

specific for NADP. It is believed that this is the principal site of reduction for the assimilated carbon dioxide in photosynthesis, although some objections must be considered later. The thylakoid electron transport system can be said to terminate with the reduction of NADP. This again shows a formal analogy with the mitochondrion, since in each case electron transport is between oxygen and a nicotinamide coenzyme. The enzyme responsible for this reduction is NADP reductase, the flavoprotein described on page 57. This flavoprotein will accept electrons from ferredoxin, and this in turn is reduced by thylakoid preparations in the light. The reduction of ferredoxin alone is not easy to demonstrate if oxygen is evolved, or if air is present, as the ferredoxin is reoxidized, but an artificial electron-donor, ascorbate plus dichlorophenol indophenol, can be used anaerobically. There seems little doubt that the final sequence of the electron transport pathway is represented by: reaction centre I: X_I: ferredoxin, NADP reductase and NADP (Figure 4.9). The symbol X_I is only to allow for a hypothetical intermediate: the evidence for it will be discussed later.

4.54 Two light reactions

The reduction of Hill oxidants does not always correspond with the reduction of NADP by the above sequence, and for various reasons, set out and discussed later, the idea of two light reactions in series has been introduced. That labelled System I reduces NADP as described above, and generates an oxidant, which can be labelled Y_I. System II oxidizes water and generates a reductant, which we have already labelled X_{II}, and which is oxidized by Hill reagents. It is necessary to suppose that X_{II} can be oxidized by Y_I either directly or via a chain of redox intermediaries. This scheme is

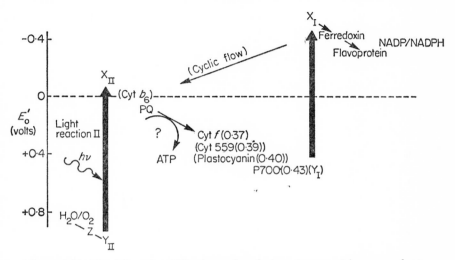

Figure 4.10. The 'zigzag' or 'Z' scheme for electron transport in green plants

summarized in Figure 4.10. It is tempting to try to place all the pathways on a common diagram, so as to produce a 'metabolic map' for electron transport. At present the combined uncertainties of each path tend to arouse confusion. Thus there is no guarantee that the site of ATP formation is the same for cyclic and non-cyclic pathways. Figure 4.10 also shows in parenthesis, positions of redox materials which are at present subject to criticism. Evidence which affects the placing and position of each redox material on a figure such as Figure 4.10 (if such a figure represents the state of affairs at all) will be reviewed later.

4.55 *The bacterial system*

Illumination of the chromatophores of *Rhodospirillum rubrum*, a purple non-sulphur bacterium, causes the oxidation of a cytochrome, C422. This reaction takes place at temperatures of 77°K indicating that it is an electron movement, not involving movements in atomic nuclei, and hence that it is a primary photoreaction. Very little else can be said concerning the electron transport system. Bacterial cells do not generate oxygen, and the photosynthetic process is anaerobic. ATP is produced, and this must be directly coupled to the photosynthetic electron transport. When provided with the usual coupling agents, chromatophores carry out cyclic photophosphorylation. Figure 4.11 is one representation of electron transport in *Chromatium*, a purple sulphur bacterium.

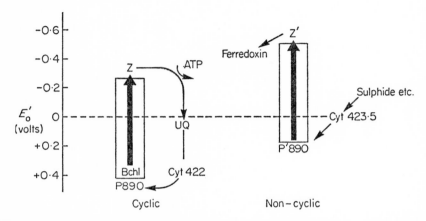

Figure 4.11. Illustrating hypothesis for electron transport in the photosynthetic bacterium *Chromatium*, after Hind and Olson (see Fig. 8.12)

Clearly, since photosynthetic bacteria assimilate carbon using light energy, non-cyclic electron transport must be taking place. The point has been made however that given a supply of ATP from the cyclic process, the non-cyclic electron transport might be a 'dark' process similar to the reduction of NAD by succinate in mitochondria. In mitochondria the process is known as reverse electron transport.

Assuming that the non-cyclic electron transport system is analogous to that in the chloroplast, then measurements of the rate of reaction with light intensity indicate that at least two quanta are required per electron. The bacterial system operates between much closer values of redox potential, and it is not easy to see why two quanta are necessary. ATP is formed during the non-cyclic process. The status of this apparent non-cyclic photophosphorylation will be discussed in section II.

The relations between the electron transport pathway and the carbon metabolism of these bacteria are capable of great variation. Hydrogen donors encountered are sulphide, thiosulphate, ethanol, butyrate, acetate, NADH and hydrogen. Hydrogen acceptors are NAD(P), nitrate, nitrite, possibly dehydrogenase reactions other than via NAD, and, in addition, hydrogen can be released as hydrogen gas. Table 4.1 indicates the distribution of the donors with the systematic arrangement of the photosynthetic bacteria. It appears that whether a bacterium growing by photosynthesis takes up or gives out carbon dioxide depends to a large extent on the hydrogen and carbon sources in the environment. This is the justification for describing nutrition in terms of chemotrophy and phototropy, rather than heterotrophy and autotrophy.

Table 4.1. Hydrogen donors in bacterial photosynthesis.

Group	Metabolism	Donors	Reductive cycles	
			Citrate	Pentose
Green Sulphur Bacteria (Chlorobacteriaceae)				
Chlorobium	autotrophic	inorganic S-cpd	+	+
Chloropseudomonas ethylicum	myxotrophic	organic-C_2 (CO_2 required)	+	−
Purple Sulphur Bacteria (Thiorhodaceae)				
Chromatium	autotrophic	S^{2-}, H_2		
	photoheterotrophic	organic	−	+
Purple Non-sulphur Bacteria (Athiorhodaceae)				
Rhodomicrobium	photoheterotrophic	organic	−	+
Rhodospirillum *Rhodopseudomonas*	autotrophic photoheterotrophic	H_2, (some) S^{2-} organic	+	+

Note (1) In the heterotrophic cases CO_2 may be either evolved or consumed (assimilated) depending on the particular substrate and the metabolic needs of the cells.
(2) *Rhodospirillum* and *Rhodopseudomonas*, when living chemotrophically, possess an (aerobic) oxidative citrate cycle.

4.56 Energy considerations

According to the scheme of Figure 4.10, about 0·8 eV of the 1·8 eV available per quantum is trapped as oxidoreduction energy. Overall, however, the total energy stored is represented by the standard potential difference between oxygen and NADPH, equal to 1·2 eV per electron. Half a molecule of ATP per electron adds 0·17 eV,* so that the energy fixed out of two quanta (3·6 eV) amounts to approximately 1·4 eV. This represents a high efficiency of energy conversion (39%).

* This figure is calculated from the formula $\Delta F_0 = nFE$ taking ΔF_0 as 8000 cal mole^{-1} and $n = 2$. Since the phosphorylation of ADP can proceed well past the standard condition, the actual free energy change, and hence the equivalent calculated redox potential difference may well be twice the above figures.

4.6 The formation of ATP

To approach the problem of the 'coupling' between electron transport and the formation of ATP, it is convenient to consider oxidative phosphorylation in the mitochondrion and photophosphorylation in the chloroplast as the same process in different locations; the problem and the method of attack are the same in each case. The organelle can be treated as a 'black box'; all conceivable models are set out, and their predicted properties compared with the behaviour of the black box under the various available experimental

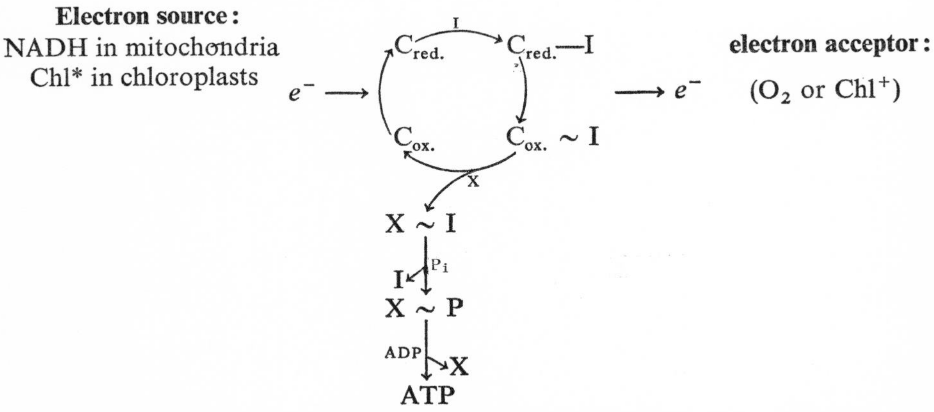

Electron source:
NADH in mitochondria
Chl* in chloroplasts

$e^- \longrightarrow$

$C_{red.} \longrightarrow C_{red.} — I$

$C_{ox.}$ $C_{ox.} \sim I$

$\longrightarrow e^-$ **electron acceptor:**

$(O_2$ or $Chl^+)$

$X \sim I$

$I \qquad P_i$

$X \sim P$

$ADP \qquad X$

ATP

Figure 4.12. Representation of the 'chemical intermediate' hypothesis for the coupling of ATP formation (phosphorylation) to electron transport in either mitochondria or photosynthetic systems

configurations. This approach does not ignore the possibility that different solutions may be found in the end for the two organelles.

Two kinds of model are prominent, and it is not possible at present to recommend either before the other. The older model, the chemical intermediate hypothesis, envisages that one of the redox materials at a given phosphorylation site forms a chemical bond with a high negative free energy of hydrolysis, which leads by transfer reactions with energy conservation to 'high-energy' phosphorylated compounds which transfer their phosphate to ADP in a kinase reaction. There is in this view a 'transferase chain' in addition to the oxidoreductase chain of electron transport. Components of such a chain could be called coupling factors, without which ATP formation could not occur. 'Uncoupling' which is observed as unrestricted electron transport with no phosphorylation taking place, would be interpreted as a hydrolytic breakdown of one of the substrates of the transferase chain (see Figure 4.12).

The alternative, newer hypothesis is that put forward by Mitchell, and termed the chemi-osmotic hypothesis. In this view electron carriers are so located in the membrane material that hydrogen ions are pumped from one face to the other. (One should remember that the membranes of mitochondrial cristae and chloroplast thylakoids where electron transport

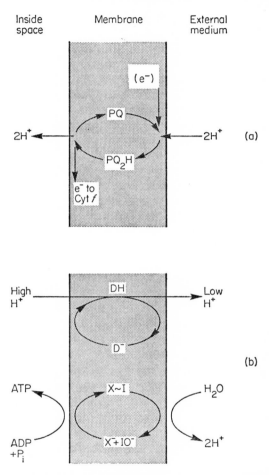

Figure 4.13. The 'chemiosmotic' hypothesis for coupling phosphorylation to electron transport. In the upper diagram electron transport causes an accumulation of H^+ ions on one side of a membrane, which may exchange with say K^+ to produce an electric field, the combined effect being to operate the system shown in the lower diagram, which may be thought of as an ATP-driven H^+-pump running in reverse

and phosphorylation occur are seen to be much thicker than other cell membranes in sections under the electron microscope.) This is brought about by the alternation of carriers such as cytochromes, which carry electrons, and those such as flavins, which carry hydrogen atoms. Hydrogen ions are presumed to be taken up at one face by the reduction of the second by the first type, and released at the other face by the reduction of a further member of the first by the second type (see Figure 4.13).

The model continues by postulating substances that condense on the acid side, migrate across the membrane (conveying the hydrogen ions down the gradient) and then dissociate again by phosphorylase and kinase reactions. The coupling factors in this case are the systems carrying out the final process. Uncoupling on this theory is a process which interferes with the integrity of the membrane, or allows recombination of the separated ions other than through the coupling site, or (in the case of ammonium ions and primary amines) neutralizes the hydrogen-ion gradient.

On both hypotheses, the final enzymes should have ATPase activity, and particles are known (see Plate 9 facing p. 84) from the membrane surface of both mitochondria and chloroplasts which have such activity, and have the properties of coupling factors.

4.7 van Niel's Hypothesis

A major step in the comparison of photosynthetic systems was made by van Niel in 1933. He rationalized the great variety of chemical reactions carried out in bacterial photosynthesis and in the green-plant process involving oxygen production by supposing that the initial step of photosynthesis was the splitting of water. This photolysis of water led to the formation of an oxidant and a reductant. The difference between plants and bacteria on this basis was that green plants had the means to release oxygen from the oxidant as gas, while the anaerobic bacteria released it by combination with an environmental reductant. This hypothesis, represented by the diagram in Figure 4.14, stimulated research for some thirty years. However the elucidation of electron transport chains has shown now that separation of charge explains the same observations and is easier to understand as the primary photochemical process. When as must usually happen the elements of water enter into the half-cell reactions at either end of the electron transport pathway, this can be regarded as an ionic reaction with either hydrogen or hydroxyl ions. The formation of these is a fast spontaneous reaction, and in no sense a photolysis. Nevertheless, the simultaneous removal of hydrogen and hydroxyl ions by the two electron transport termini requires the same

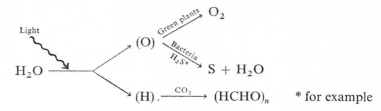

Figure 4.14. Van Niel's hypothesis, that in all photosynthetic processes water is split to give an oxidant and a reductant

energy and has the same overall formulation as the hypothetical photolysis of a water molecule. A modern translation of van Niel's hypothesis would be that photosynthesis involves a *separation of charge* leading to the formation of an oxidant and a reductant.

4.8 Summary

The thylakoids of the green plant in conjunction with ferredoxin evolve oxygen and reduce NADP in light, and a widely accepted possible mechanism is epresented in Figure 4.10. By this means ATP and NADPH are supplied to the stroma enzymes, probably in equimolar proportions. This is the non-cyclic process of electron transport and photophosphorylation.

In the cyclic process (Figure 4.7) chloroplasts or chloroplast fragments supplied with ferredoxin form ATP in the light. By the two processes together the thylakoids can supply ATP in a higher proportion to NADPH.

Both processes probably occur in the bacteria.

References

Klingenberg, M. (1968). The respiratory chain. In T. P. Singer (Ed.), *Biologica Oxidations*, Interscience, New York, p. 3.

Mahler, H. R. and E. H. Cordes (1966). *Biological Chemistry*, Harper and Row, New York, p. 208.

Rabinowitch, E. (1945, 1951, 1956). *Photosynthesis and Related Processes*, Vols. 1, 2 and 3, Interscience, New York.

Smith, L. (1968). The respiratory chain system of bacteria. In T. P. Singer (Ed.), *Biological Oxidations*, Interscience, New York, p. 55.

Strittmatter, P. (1968). Microsomal electron transport. In T. P. Singer (Ed.), *Biological Oxidations*, Interscience, New York, p. 171.

Suggested further reading

Hill, R. (1965). The Biochemists' Green Mansions; the photosynthetic electron-transport chain in plants. In P. N. Campbell and G. D. Greville (Eds.), *Essays in Biochemistry*, Vol. 1, Academic Press, London, p. 121.

Arnon, D. I. (1968). Electron transport and photophosphorylation in chloroplasts. In T. P. Singer (Ed.), *Biological Oxidations*, Interscience, New York, p. 123.

Glasstone, S. and D. Lewis (1960). *Elements of Physical Chemistry*, 2nd ed., Chapter 13, Macmillan, London.

CHAPTER 5

The path of carbon

The model so far presented separates photosynthesis into 'light' and 'dark' reactions, the light reactions being the pigment-driven electron transport pathways (Chapter 4) which result in the formation of NADPH and ATP. The dark reactions now to be described involve a series of enzymes in the stroma of the chloroplast which use the ATP and NADPH for the reduction of carbon dioxide, so that ATP and NADPH are the only link between the thylakoid and stroma. Similar relationships are supposed to hold between both the chromatophores of photosynthetic bacteria and the thylakoids of blue–green algae and their respective cytoplasms. While there are still some difficulties in this model, it has great strength since there can be no doubt that thylakoid preparations can be made to produce ATP and NADPH, and also that enzymes do exist in the stroma which given these materials will reduce carbon dioxide.

In this section we shall consider sequences of enzyme-catalysed reactions that have been put together as 'metabolic pathways', to explain the formation of cell material from the carbon sources used by photosynthetic organisms. In the green plants carbon dioxide is usually used to make carbohydrate. (In some cases other products may predominate, such as fat or glycolate derivatives; this is particularly important in algae. These pathways are left to Chapter 10 for discussion, as is also an alternative route to carbohydrate currently being elucidated by Hatch and Slack.) We shall examine here, with the green plants, some schemes for the very varied metabolism found among the photosynthetic bacteria.

5.1 The incorporation of carbon dioxide in the chloroplast

This pathway for the incorporation of carbon dioxide and its reduction to carbohydrate was discovered by the group at Berkeley led by Calvin. It is cyclic in operation, and has been variously termed the 'Calvin Cycle', the 'Calvin–Benson Cycle' and the 'Photosynthetic Carbon Cycle'. However, although it may well represent the principal pathway for photosynthesis in most green plants under 'normal conditions' ('normal conditions' referring to open-air, temperate-zone vegetation in a maritime climate in sunlight!) other cyclic pathways do exist in bacteria and some green plants. We will use the term 'Reductive Pentose Cycle' for the sequence of reactions set out in Figure 5.1.

Benson studied this pathway by allowing cultures of the green alga *Chlorella* to incorporate radioactive (^{14}C) carbon dioxide by photosynthesis for limited periods of time. The cells were then killed and their contents separated and analysed by the technique of *chromatography*, usually in two dimensions on paper, so that the materials are distributed in 'spots', the coordinates of which are characteristic of the individual substances. The spots are revealed either by staining with chemical reagents, or, in the case of radioactive materials, by clamping an X-ray film against the paper for a time, the radiation from the spots causing them to appear as dark areas when the film is developed. This process is *radioautography*. It was found that the first material to contain radioactivity was 3-phospho-D-glyceric acid (PGA); longer periods of photosynthesis resulted in a greater number of materials becoming radioactive, from which a time sequence could be deduced. Plate 10 (facing page 84) is a radioautograph of such a preparation. This work depended on the then very new technique of chromatography, and the new isotope ^{14}C. These workers and others, including the group led by Arnon, also at Berkeley, showed that the cells contained enzymes that would catalyse the reactions that had been postulated. The elucidation of the pathway was a great achievement, recognized by the award of a Nobel Prize to Calvin in 1961.

Figure 5.1 sets out the reactions of the reductive pentose cycle so as to show the overall formation of carbohydrate in the formation of D-glyceraldehyde-3-phosphate from 3 molecules of carbon dioxide. Glyceraldehyde-3-phosphate (G3P)

$$
\begin{array}{l}
\text{CHO} \\
| \\
\text{HCOH} \\
| \\
\text{CH}_2\text{O}\textcircled{P}
\end{array}
$$

ⓟ represents —PO(OH)$_2$
(orthophosphate)

G3P

is a phosphate ester of a *triose;* trioses are monosaccharide sugars of three carbon atoms. The carbohydrates involved in the reductive pentose cycle include tetroses (4 carbons), pentoses (5), hexoses (6) and heptoses (7). Although G3P is an important biochemical material, it is not present in large quantities, and the product of photosynthesis that is usually observed is sucrose (made of one molecule each of the hexoses glucose and fructose) or

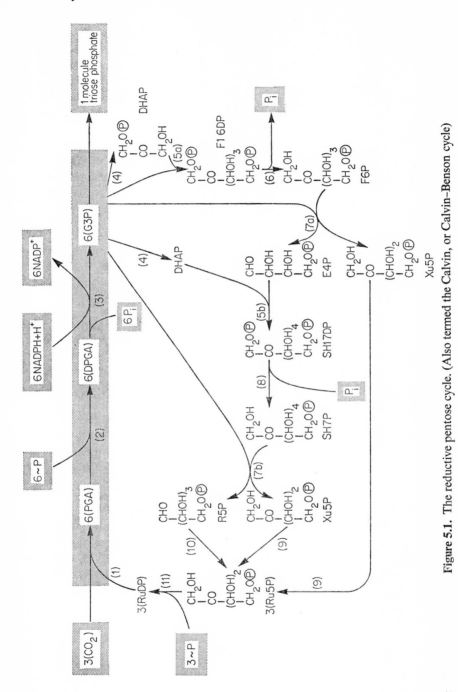

Figure 5.1. The reductive pentose cycle. (Also termed the Calvin, or Calvin–Benson cycle)

starch (a large molecule formed by condensation of many glucose units). The interconversion of carbohydrates is, however, a metabolic process occuring in most organisms, including the classical materials of biochemical study: yeast cells, muscle and liver tissues. Once carbohydrate has been fed into this system in almost any form it can be easily converted to any other form required by the cell at the time. It is therefore convenient to regard the specifically photosynthetic part of the process as terminated with the production of carbohydrate, here triose phosphate.

The reactions of the cycle are identified both by name and, in the text, by their reference numbers in the Enzyme Commission scheme of classification. The main features of this scheme are as follows. There are four numbers in a reference, the first denoting which of six classes of reaction the enzyme catalyses. In class EC.1, the *oxidoreductases* catalyse reactions of the type

$$AH_2 + B \xrightarrow{\text{EC.1}} A + BH_2$$

In the cycle, an example is the reduction of diphosphoglyceric acid (DPGA) to G3P by the enzyme EC.1.2.1.13. The second number here indicates that the substrate *acid* is reduced to *aldehyde* and the third that the hydrogen is carried by a nicotinamide coenzyme, in this case NADP:

$$NADPH + \begin{matrix} COO\circledP \\ | \\ HCOH \\ | \\ CH_2O\circledP \end{matrix} \xrightarrow{\text{EC.1.2.1.13}} \begin{matrix} CHO \\ | \\ HCOH \\ | \\ CH_2O\circledP \end{matrix} + NADP^+ + H^+ + P_i$$

The fourth number is a specific serial number.

The second class, of transferring enzymes (*transferases*) may be represented by the scheme:

$$AB + C \xrightarrow{\text{EC.2}} A + BC$$

The second and third numbers describe the nature of the group transferred ('B') and sometimes the nature of the acceptor ('C'). The fourth number as always is the specific serial reference.

The third type of reaction is *hydrolysis* which is the breaking of a molecule by the addition of the elements of water, thus:

$$AB + H_2O \xrightarrow{\text{EC.3}} AH + BOH$$

Enzymes of this group are the *hydrolases*, and in the reductive pentose cycle they are represented by *phosphatases*, where the group 'B' in the above equation is the phosphate group that we have shown by the symbol $\circledP$.

Class EC.4, the lyases, nominally break molecules into two products:

$$AB \rightarrow A + B$$

but there are two points to make. First, the reaction may be found going the opposite way from the nominal direction above, and in the examples of the *aldolase* enzymes of the cycle, the reactions are indeed reversible. Secondly, the key enzyme of the cycle that adds carbon dioxide to a pentose, producing two molecules of PGA, is classified as a lyase, which might be thought surprising.

Class EC.5 contains the *isomerases*, which interconvert pairs of chemical isomers:

$$ABC \xrightarrow{EC.5} ACB$$

We may distinguish *epimerases* (EC.5.1) which invert the stereochemical configuration at a specific point, from the *intramolecular oxidereductases* (EC.5.3) which here interconvert aldose and ketose sugars:

CHO   EC.5.3.1.1   CH$_2$OH
| ⇌ |
HCOH   CO
| |
CH$_2$O℗   CH$_2$O℗

G3P   Dihydroxyacetone
(aldotriose   phosphate
phosphate)   (ketotriose phosphate)

CHO   CH$_2$OH
| |
HCOH   CO
| EC.5.3.1.6 |
HCOH   ⇌   HCOH
| |
HCOH   HCOH
| |
CH$_2$O℗   CH$_2$O℗

Ribose-5-phosphate   Ribulose-5-phosphate
(aldopentose   (ketopentose phosphate)
phosphate)

In general, ketosugars are named after a parent aldosugar, inserting the letters -ul-, thus ribulose is the ketosugar related to ribose (and also arabinose). An example of an epimerase reaction is the conversion of xylulose to ribulose phosphate by epimerization at the 3-position:

1   CH$_2$OH   CH$_2$OH   1
| |
2   CO   CO   2
| EC.5.1.3.4 |
3   HOCH   ⇌   HCOH   3
| |
4   HCOH   HCOH   4
| |
5   CH$_2$O℗   CH$_2$O℗   5

The sixth class, the ligases, represented by the general equation

$$A + B + ATP \xrightarrow{\text{EC.6}} AB + ADP + P_i$$

are not found in the reductive pentose cycle.

5.11 Description of the reductive pentose cycle

Referring to Figure 5.1, we shall commence with the entry of carbon dioxide: the enzyme *carboxydismutase*, also known as *ribulosediphosphate carboxylase*, EC.4.1.1.39, carries out the reaction:

$$
\begin{array}{ll}
\mathrm{CH_2O\textcircled{P}} & \mathrm{CH_2O\textcircled{P}} \\
| & | \\
\mathrm{CO + CO_2} & \mathrm{HOCH{-}COOH} \\
| & \\
\mathrm{HCOH} & \mathrm{COOH} \\
| & | \\
\mathrm{HCOH} & \mathrm{HCOH} \\
| & | \\
\mathrm{CH_2O\textcircled{P}} & \mathrm{CH_2O\textcircled{P}}
\end{array}
$$

$$\mathrm{RuDP + CO_2} \xrightarrow{\text{Mg}^{2+}} 2\mathrm{PGA} \tag{1}$$

(RuDP stands for ribulosediphosphate)

In this reaction one would suspect a six-carbon intermediate, but in spite of much searching, none has been found. It is a matter of some concern that, after making allowance for its abundance, isolated preparations of the enzyme are not sufficiently active to explain the rate of carbon dioxide uptake by plants under field conditions. It constitutes nearly half of all leaf protein and is often referred to as 'Fraction-I' protein.

The second step (from our arbitrary starting point) is one of the two sites where ATP is required; ATP is a necessary source of energy, and is provided by the process of photophosphorylation in the thylakoids (section 4.6). Here the terminal phosphate group of ATP is transferred to the carboxyl group (carbon-1) of PGA to make a 'mixed anhydride'. As in the case of the

$$
\begin{array}{ll}
\mathrm{COOH} & \mathrm{CO.O\textcircled{P}} \\
| & | \\
\mathrm{HCOH} \quad + ATP \xrightarrow[\text{EC.2.7.2.3}]{\text{Carboxydismutase}} ADP + \mathrm{HCOH} \\
| & | \\
\mathrm{CH_2O\textcircled{P}} & \mathrm{CH_2O\textcircled{P}}
\end{array}
\tag{2}
$$

PGA

1,3 diphospho-
glyceric acid
(1,3 DPGA)

glycolysis pathway where the same enzyme acts, the equilibrium is in favour of the synthesis of ATP. However the impetus of the cycle is maintained by the overall energy change, which is large enough to make the cycle as a whole effectively irreversible.

1,3-Diphosphoglyceric acid (1,3DPGA) is a substrate for the enzyme *triosephosphate dehydrogenase* (or more formally D-glyceraldehyde-3-phosphate dehydrogenase) EC.1.2.1.13 acting in the reverse direction to its name. This step provides the site of entry for the third and last 'raw material' of the cycle: the reduced coenzyme NADPH. It may be noted that the corresponding enzyme in the glycolysis pathway, EC.1.2.1.12, uses the coenzyme NAD. In a plant cell the former occurs inside the chloroplast, the latter outside in the cell cytoplasm. The NADP-coupled enzyme is much more difficult to isolate. The product of the reaction (3), glyceraldehyde-3-phosphate, is the final product that we have set ourselves to produce; however, we have only accounted for the uptake of one molecule of carbon dioxide, and have not explained the origin of the ribulosediphosphate used in the first step. The rest of the cycle therefore will take five-sixths of the triosephosphate and reconvert it to ribulose-1,5-diphosphate. This process will consume the remaining requirement of ATP. The reactions of this regenerative part of the cycle are almost exactly those of the pentose phosphate pathway in animals and bacteria, which is why it is instructive to refer to the overall process as the 'reductive' pentose cycle.

The reactions catalysed by *triosephosphate isomerase* EC.5.3.1.1, *aldolase* EC.4.1.2.13 and fructose *diphosphatase* EC.3.1.3.11 (4, 5 and 6 respectively) are those of glucogenesis; transketolase EC.2.2.1.1. is apparently the same as in the pentose phosphate pathway of animals. Transketolase occurs twice in Figure 5.1, in the reactions

$$
\begin{array}{c}
\text{CH}_2\text{OH} \\
|\\
\text{CO} \\
|\\
\text{HOCH} \\
|\\
\text{HCOH} \\
|\\
\text{HCOH} \\
|\\
\text{CH}_2\text{P}
\end{array}
+
\begin{array}{c}
\text{CHO} \\
|\\
\text{HCOH} \\
|\\
\text{CH}_2\text{OP}
\end{array}
\underset{\text{Mg}^{2+}}{\overset{\text{TPP}}{\rightleftharpoons}}
\begin{array}{c}
\text{CHO} \\
|\\
\text{HCOH} \\
|\\
\text{HCOH} \\
|\\
\text{CH}_2\text{OP}
\end{array}
+
\begin{array}{c}
\text{CH}_2\text{OH} \\
|\\
\text{CO} \\
|\\
\text{HOCH} \\
|\\
\text{HCOH} \\
|\\
\text{CH}_2\text{OP}
\end{array}
\qquad (7a)
$$

| Fructose-6 phosphate (F6P) | G3P | Erythrose-4-phosphate (E4P) | Xylulose-5-phosphate (Xu5P) |

$$
\begin{array}{l}
\text{CH}_2\text{OH} \\
|\\
\text{CO} \\
|\\
\text{HOCH} \\
|\\
\text{HCOH} \\
|\\
\text{HCOH} \\
|\\
\text{HCOH} \\
|\\
\text{CH}_2\text{O\textcircled{P}}
\end{array}
+
\begin{array}{l}
\text{CHO} \\
|\\
\text{HCOH} \\
|\\
\text{CH}_2\text{O\textcircled{P}}
\end{array}
\underset{\text{Mg}^{2+}}{\overset{\text{TPP}}{\rightleftharpoons}}
\begin{array}{l}
\text{CHO} \\
|\\
\text{HCOH} \\
|\\
\text{HCOH} \\
|\\
\text{HCOH} \\
|\\
\text{CH}_2\text{O\textcircled{P}}
\end{array}
+
\begin{array}{l}
\text{CH}_2\text{OH} \\
|\\
\text{CO} \\
|\\
\text{HOCH} \\
|\\
\text{HCOH} \\
|\\
\text{CH}_2\text{O\textcircled{P}}
\end{array}
\qquad \text{(7b)}
$$

Sedoheptulose-7   G3P   Ribose-5-   X5P
phosphate   phosphate
(SH7P)   (R5P)

It transfers the carbons 1 and 2 (as a unit) from a ketose to an aldose acceptor; since the remains of the donor is now an aldose (with 2 carbons less) and the acceptor has become a ketose (with 2 carbons more) the reaction is reversible.

Thiamine pyrophosphate acts as the C_2-carrier, and it appears that the pool of TPP-glycolaldehyde adduct is common to the two reactions. *Glycolate* formed under some conditions may originate from this pool: this will be discussed later (Chapter 10).

The enzyme *aldolase*, EC.4.2.1.13 which adds dihydroxyacetone phosphate (DHAP) to an aldehyde thus forming a ketose-1-phosphate, also has two sites of action, (5a) where the aldehyde is G3P, making fructose-1,6-diphosphate (Fl6DP), and another (5b), that may possibly be a different form of the enzyme, that makes sedoheptulose-1,7-diphosphate (SH17DP) using erythrose-4-phosphate (E4P):

$$
\begin{array}{l}
\text{CH}_2\text{O\textcircled{P}} \\
|\\
\text{CO} \\
|\\
\text{CH}_2\text{OH}
\end{array}
+
\begin{array}{l}
\text{CHO} \\
|\\
\text{HCOH} \\
|\\
\text{HCOH} \\
|\\
\text{CH}_2\text{O\textcircled{P}}
\end{array}
\xrightarrow{\text{Aldolase}}
\begin{array}{l}
\text{CH}_2\text{O\textcircled{P}} \\
|\\
\text{CO} \\
|\\
\text{HOCH} \\
|\\
\text{HCOH} \\
|\\
\text{HCOH} \\
|\\
\text{HCOH} \\
|\\
\text{CH}_2\text{O\textcircled{P}}
\end{array}
\qquad \text{(5b)}
$$

DHAP   E4P   SH17DP

There is a specific hydrolase, sedoheptulose diphosphatase, that removes the 1-phosphate, leaving SH7P (8).

The pentose phosphates Xu5P and R5P are both converted (9, 10) to Ru5P by means of the enzymes ribulosephosphate-3-epimerase EC.5.1.3.4 and ribosephosphate isomerase EC.5.3.1.6 respectively. Finally the ribulose-5-phosphate is phosphorylated (11) at the 1-position by ATP (completing the account of the 'raw materials') giving RuDP and completing the cycle.

In summary, the reductive pentose cycle has been described as the uptake of three molecules of carbon dioxide by three molecules of RuDP giving six PGA, of which one is a product; the remaining five PGA molecules (totalling fifteen carbon atoms) are rearranged into three molecules of RuDP. During this cycle nine molecules of ATP and six of NADPH are consumed. This can be represented in one line:

$$3RuDP + 3CO_2 \xrightarrow{\text{9ATP + 6NADPH}} G3P + 3RuDP$$

The pentose phosphate pathway in animals involves the enzyme *transaldolase*, avoiding the formation of sedoheptulose-1,7-diphosphate. However, since transaldolase has been shown to exist only in a few plant species, and in these only in small quantities, it has been concluded that it is not involved in the reactions of the reductive pentose phosphate cycle.

Although many of the enzymes are common to pathways in animal tissues, studies in the latter indicate an elaborate series of controls, which may differ from source to source. The classification of chloroplast enzymes into general categories may conceal important differences. The discovery of allosteric and other regulatory effects in the enzymes of, say, glycolysis has led to some remarkable conclusions, and it can be expected that enzymes of the above cycle will be found to possess a similar degree of control.

5.2 Carbon metabolism in the phototrophic bacteria

Chromatophores from all three main groups of photosynthetic bacteria carry out a cyclic photophosphorylation process. Non-cyclic electron transport has been observed, ending with the reduction of ferredoxin, or NAD (not apparently NADP); the reduction of NAD is independent of ferredoxin. The question whether this non-cyclic process is directly light-driven can be left until a later chapter; for the time being, it should be remembered that the membrane material of photosynthetic bacteria when illuminated forms NADH and ATP.

5.21 Incorporation of organic materials

While some bacteria can photosynthetically reduce carbon dioxide with an inorganic hydrogen donor, others are able to use organic materials as the source of carbon and hydrogen. The second case will be considered first. All three groups of bacteria carry out such a process, although the biochemical means differ. There are two separate dark reactions, one which oxidizes the substrate, producing reducing equivalents, and the other reductive incorporating the substrate into cell constituents. This kind of process is known in chemistry as a 'disproportionation' reaction, (Figure 5.2), and can occur spontaneously in a few cases. In a photosynthetic

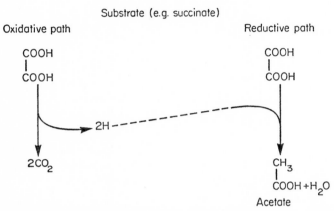

Figure 5.2. An oxidative and a reductive pathway coupled to form a 'disproportionation' reaction

context the substrates available do not undergo the reactions spontaneously (which is the same as saying that they are close to a state of chemical equilibrium), and energy must be supplied to the system to cause it to run in the direction indicated. This energy of course comes from light, and may be mediated in two ways. First, the excitation energy from the light may be used to drive a non-cyclic electron transport system, which takes electrons at one potential, and drives them to a much lower (more reducing) potential (Figure 5.3). (It should be stressed that electrons are not available from water, in contrast to non-cyclic electron flow in green plants.) There is some doubt whether non-cyclic, light driven electron flow actually takes place in bacteria; the same result could be obtained by the second mediation system, using ATP provided by cyclic, light-driven phosphorylation. By coupling a reaction to the splitting of ATP, the total change in free energy may be sufficiently negative to carry it on. ATP could be applied to the system of Figure 5.2 at two possible sites (Figure 5.4).

The green sulphur bacterium *Chloropseudomonas ethylicum* oxidizes acetate by the citric acid cycle (an outline of this is shown in Figure 5.5), and

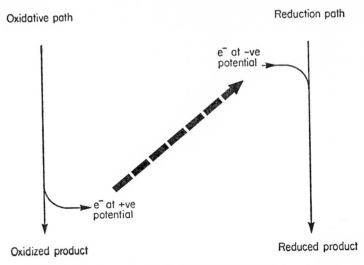

Figure 5.3. The system of Fig. 5.2 driven by light energy.

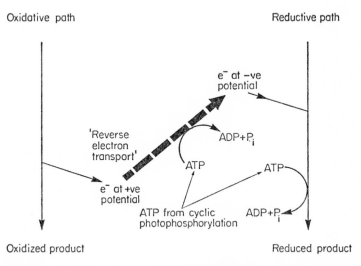

Figure 5.4. The system of Fig. 5.3 driven by ATP from either cyclic photophosphorylation or elsewhere

it possesses the modification to this pathway known as the glyoxylate pathway which enables it to metabolize acetate avoiding the immediate loss of two carbon atoms as carbon dioxide (see Figure 5.6). The reducing equivalents available from the oxidation path via the citric acid cycle, with ATP, can be used to reduce the organic acids shown to materials such as fats, carbohydrates and aminoacids.

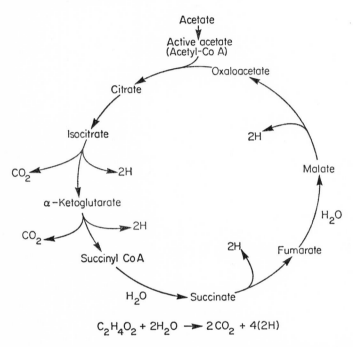

$$C_2H_4O_2 + 2H_2O \longrightarrow 2CO_2 + 4(2H)$$

Figure 5.5. The citrate cycle common to most aerobic organisms.

5.22 *Incorporation of carbon dioxide in photosynthetic bacteria*

The purple sulphur bacterium *Chromatium* can be grown on acetate in the same way, and probably using the same metabolic pathways as *Chloropseudomonas* above. This organism is however able to grow autotrophically, reducing carbon dioxide by means of an inorganic hydrogen donor such as hydrogen sulphide. Under these circumstances the bacterial cells cease to synthesize the enzymes of the glyoxylate pathway, and begin to make those of the reductive pentose pathway (section 5.1). The switching of metabolism from one route to another by means of the selective synthesis of the necessary

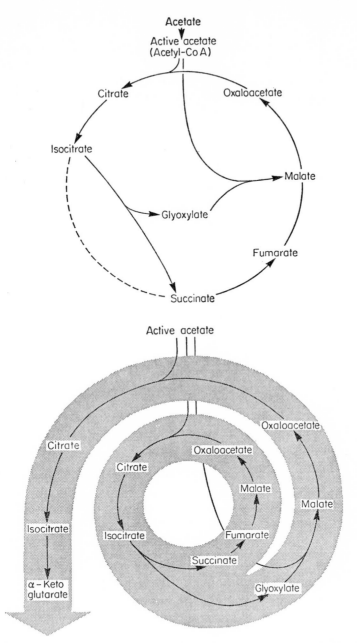

Figure 5.6. The glyoxylate pathway drawn to show its relation to the citrate cycle (a), and (b) drawn to show its operation as a cycle for synthesis.

enzymes is a common feature in the bacteria, and must account to a large extent for their great ability to survive changes in conditions.

Another species of green sulphur bacterium, *Chloropseudomonas thiosulphatophilum* possesses both the reductive pentose cycle and a new process recently discovered in the laboratory of Arnon and termed the *reductive*

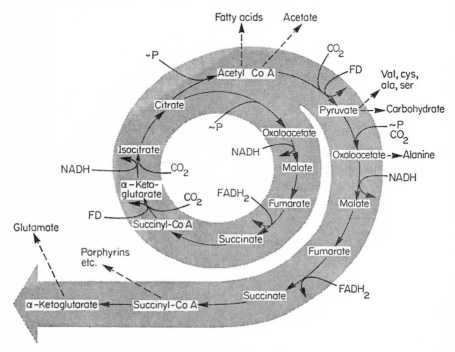

Figure 5.7. The reductive citrate cycle drawn to show its biosynthetic value. After Evans, M. C. W., B. B. Buchanan and D. I. Arnon (1966). *Proc. Natl. Acad. Sci.*, **55**, 928, with permission

citric acid cycle. This process reverses the direction of the citric acid cycle of Figure 5.5 by means of four enzymes which are adapted to reverse stages, that are normally virtually irreversible, by coupling to them energy in the form of ATP, or as the strongly-reducing agent, reduced ferredoxin. This reductive citrate cycle is shown in Figure 5.7. The key enzymes are, first, the reversal of the condensation of active acetate ('acetyl CoA') with oxaloacetate to give citrate; the reversal splits ATP to ADP and inorganic phosphate:

$$\text{Citrate} + \text{CoA} + \text{ATP} \xrightarrow{\text{EC.4.1.3.8}}$$

$$\text{Oxaloacetate} + \text{Acetyl CoA} + \text{ADP} + P_i$$

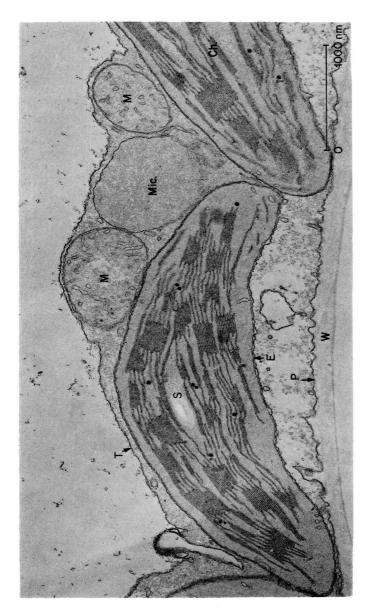

Plate 1. Shows small part of the cytoplasm of a broad bean mesophyll cell with the main organelles. Chloroplasts can be seen in contact with mitochondria and a microbody (possibly concerned in the glycolic acid metabolism of photosynthesis). The section is cut perpendicular to the membranes of the thylakoids showing that these lie predominantly parallel to the cell wall. Ch—Chloroplast. E—Envelope. M—Mitochondrion. Mic.—'Microbody' (glyoxisome? peroxisome?). P—Plasmalemma (cell outer membrane). S—Starch. T—Tonoplast membrane. Vac.—Vacuole. W—Wall (of cell). Courtesy of A. D. Greenwood, Botany Department, Imperial College, London.

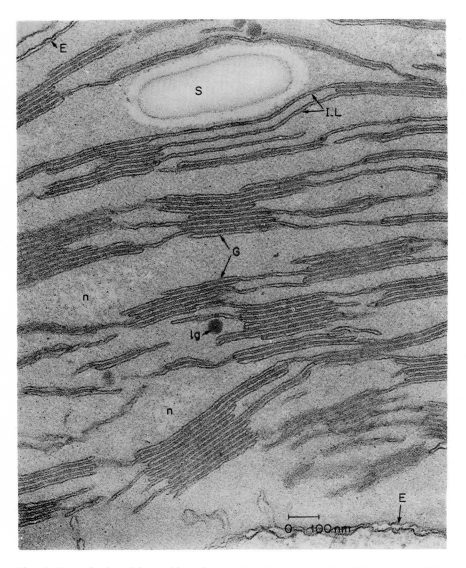

Plate 2. Part of a broad bean chloroplast to show the construction of the grana and the intergranal lamellae. E—Envelope. IL—Intergranal lamellae (intergranal thylakoids). G—Granum. lg—Lipid globule (plastoglobulus, osmiophilic globule). n—Nucleoid (DNA-containing region). S—Starch grain (note inner and outer regions). Courtesy of A. D. Greenwood, Botany Department, Imperial College, London.

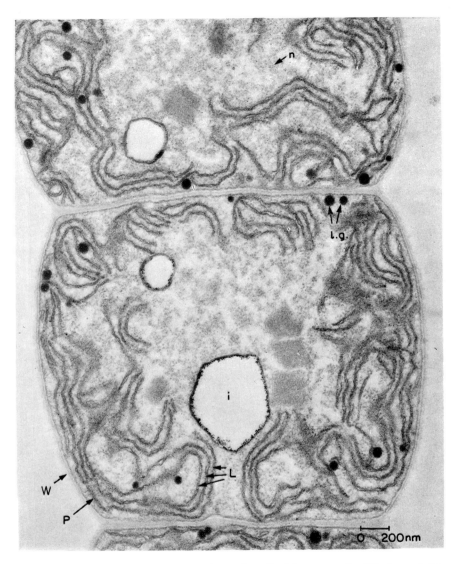

Plate 3. Electron micrograph from a thin section of a blue green alga (*Nostoc sp.*). The chlorophyll-carrying lamellae (L) or thylakoids, are situated in the general cytoplasm of the cell, mainly towards its periphery where lipid globules are also common. As in the red algae each thylakoid is separate from its neighbours with the biliprotein accessory pigments closely associated with the interior of the membranes and aggregated into phycobilisomes but these are often small in size and, as here, not easily distinguished from other cytoplasmic particles. The DNA is found as fine fibrils in the 'nucleoid' (n) regions of the cytoplasm. Other characteristic cell components are polyhedral bodies (P), probably protein, and vesciles (i) which may be contained polyphosphates lost during preparation (W)—Wall of cell. Courtesy of Dr. H. Bronwen Griffiths, Botany Department, Imperial College, London.

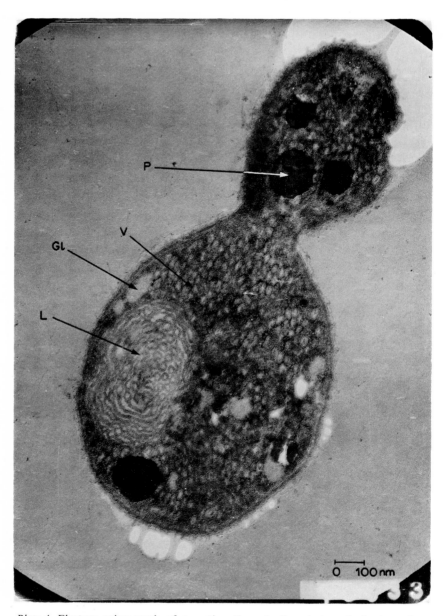

Plate 4. Electron micrograph of a section through the photosynthetic bacterium *Thiocapsa* (Thiorhodaceae). The vesicles (v) and the lamellae (L) may be regarded as corresponding to the thylakoid material of chloroplasts. (P) polymetaphosphate deposit. (Gl) glycogen deposit. Courtesy of Dr. G. Cohen-Bazire, Department of Bacteriology and Immunology, University of California, Berkeley.

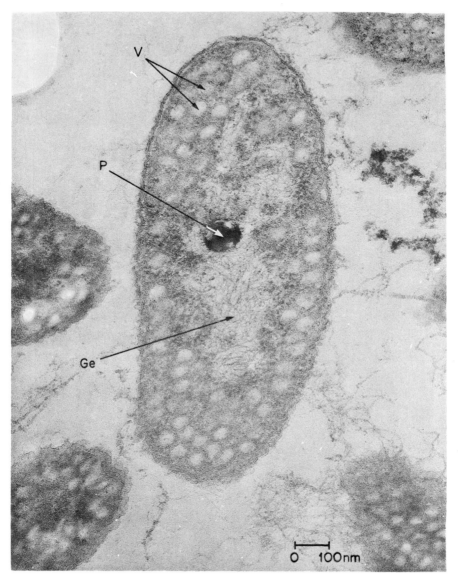

Plate 5. Section of *Rhodopseudomonas speroides* (Athiorhodaceae) grown photosynthetically at low light intensity. The photochemical apparatus is in the form of 50 nm vesicles (V). At higher light intensities lamellar material is also formed. (P) polymetaphosphate deposit. (Ge) region containing genetic material (DNA). Stained with lead hydroxide. Electron micrograph by courtesy of Dr. G. Cohen-Bazire, Department of Bacteriology and Immunology, University of California, Berkeley.

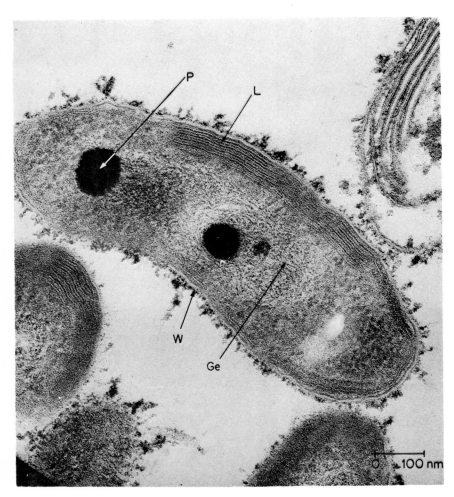

Plate 6. Rhodopseudomonas palustris (compare with Plate 5). Here lamellar material is evident (L). (W) cell wall. (Ge)—genetic material. (P)—polyphosphate deposit. Electron-micrograph by courtesy of Dr. G. Cohen-Bazire, Department of Bacteriology and Immunology, University of California, Berkeley.

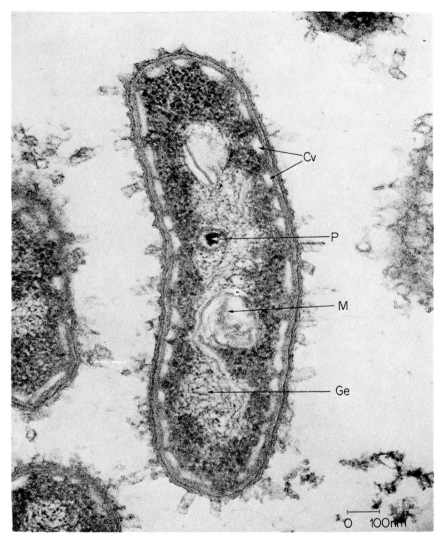

Plate 7. Chlorobium thiosulfatophilum (Chlorobacteriaceae). The vesicles (Cv) *Chlorobium* vesicles) are distributed round the periphery of the cell. (M) mesosome, an enigmatical structure, common among bacteria, which communicates with the exterior and may be involved in DNA transfer. (Ge)—genetic material. (P)—polyphosphate deposit. Micrograph by courtesy of Dr. G. Cohen-Bazire, Department of Bacteriology and Immunology, University of California, Berkeley.

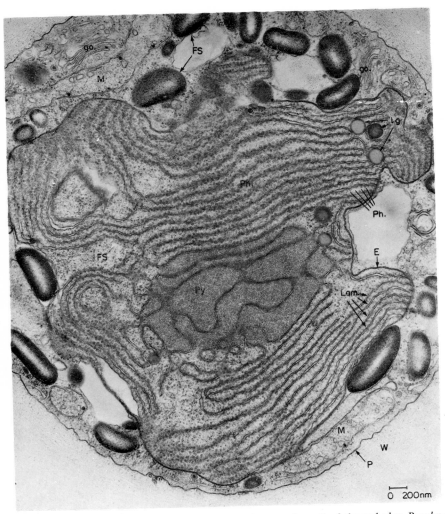

0   200nm

Plate 8. Electron micrograph of a section through part of a cell of the red alga *Porphy-ridium cruentum.* The single chloroplast which fills most of the cell is enclosed by an envelope (E) of two membranes, and contains a central pyrenoid (Py) traversed by extensions of the thylakoids. The stroma appears as a granular continuum, or matrix, supporting the thylakoid membranes (LAM) which do not form stacks or come into surface contact although they anastomose and interconnect. The surface of the thylakoid membrane bears the characteristic granules termed 'phycobilisomes' (Ph) believed to be aggregates of the biliprotein accessory pigments. Grains of floridian starch (FS), mitochondria (M), golgi bodies (go), are contained in the cytoplasm itself bounded by the plasmalemma (P) and a gelatinous sheath or wall (W). (lg)—lipid globules. Courtesy of A. D. Greenwood, Botany Department, Imperial College, London.

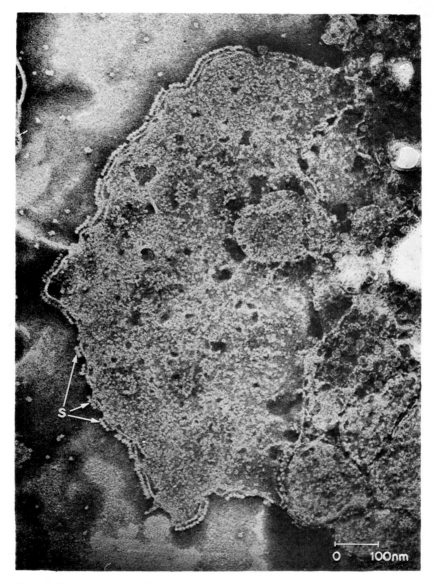

Plate 9. Fragmented thylakoids from spinach chloroplasts, dried onto a film of carbon-covered collodion and negatively contrasted by deposition of phosphotungstic acid (adjusted to pH 7). The 9 nm stalked spheres (S) are clearly visible at the edge of the large fragment. By courtesy of M. Raymond Bronchart, Department de Botanique, Universite de Liège, From C. Sivonval, (ed.) *Le Chloroplaste*, Masson, Paris, 1967, p. 55, plate IX(B), with the permission of the publisher.

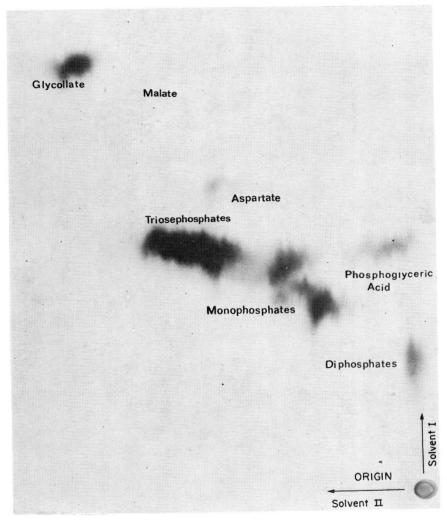

Plate 10. Reproduction of an autoradiograph of a two-dimensional chromatogram on paper of the products of photosynthesis in pea chloroplasts, using ^{14}C CO_2. The dark spot due to radioactive triosephosphate indicates that the principal pathway was the reductive pentose cycle, other intermediates of which can be seen as paler spots. The occurrence of glycolate is common, and is discussed in Chapter 10. Autoradiogram by courtesy of Professor D. A. Walker, Department of Botany, The University, Sheffield.

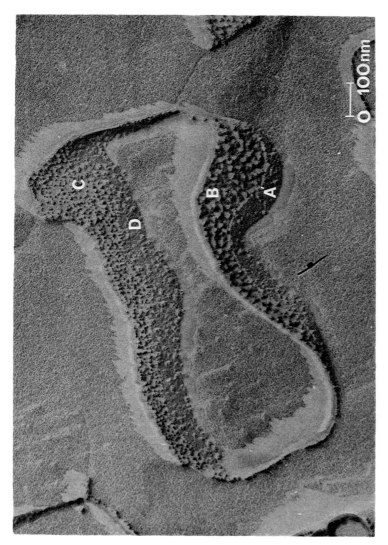

Plate 11. Deep-etched and shadowed surface of a fractured thylakoid. See Figures 7.10 and 7.11 for an explanation of the method, and an interpretation of the regions indicated here: A′ surface, B face, C face and D surface. Note that the A surface as opposed to A′ is that exposed when two appressed thylakoids are separated, a rare observation. By courtesy of Professor R. B. Park, Department of Botany, University of California, Berkeley.

This enzyme, the 'citrate cleavage enzyme' is known in animal cells, but is located away from the region where the citric acid cycle takes place.

Two enzymes reverse the 'oxidative decarboxylation of α-ketoacids' in which pyruvate and α-ketoglutarate form acetyl and succinyl CoA derivatives respectively, losing carbon dioxide and reducing the coenzyme NAD (Figure 5.7). The reversal is carried out, not with NADH, but with reduced ferredoxin. The extra 100–150 mV of reducing potential is apparently sufficient to allow the reaction to proceed. The fourth enzyme phosphorylates pyruvic acid to phosphoenolpyruvate, obtaining energy by splitting both high-energy bonds of the ATP so as to yield AMP and inorganic phosphate:

$$\begin{array}{ccc} \text{COOH} & & \text{COOH} \\ | & & | \\ \text{CO} & + \text{ATP} \longrightarrow \text{AMP} + P_i + & \text{CO}\circledP \\ | & & \| \\ \text{CH}_3 & & \text{CH}_2 \end{array}$$

(This is not related to the citric acid cycle as such but rather to a sequence of metabolic paths connecting it with the glycolysis system. In liver tissue, the overall reaction of the conversion of pyruvate to oxaloacetate takes place directly using ATP and the coenzyme biotin.) The bacteria in which this pathway has been demonstrated are all obligate anaerobes, in that they will not grow if even traces of oxygen are present. If this is due to the tendency of strongly reducing materials to be oxidized by oxygen with the production of hydrogen peroxide,

$$AH_2 + O_2 \rightarrow A + H_2O_2$$

then this raises the question, so far unanswered, of how this is avoided in the chloroplast system. (Ferredoxin has also been obtained from non-photosynthetic bacteria such as the (anaerobic) *Clostridia*, which also contain the enzymes for the formation of pyruvate and α-ketoglutarate from acetyl and succinyl CoA.)

The purple non-sulphur bacteria can live either phototrophically or chemotrophically; the latter existence requires oxygen and a respiratory substrate such as acetate and the photosynthetic pigments are no longer produced. The colour of the cells then fades almost completely as the bacteria grow. When the oxygen supply is exhausted, the cells given light reform their red pigment and the underlying bacteriochlorophyll and commence photosynthesis. The photosynthetic incorporation of acetate does not appear to involve the glyoxylate pathway, and may involve the simultaneous fixation of carbon dioxide. These bacteria, like the green and purple sulphur bacteria, contain ferredoxin, and its function in carbon assimilation is not established. However there is a process whereby hydrogen gas can be evolved from

4

substrates such as acetate, succinate, fumarate and L-malate, which are completely oxidized to carbon dioxide. For example

$$CH_3 \cdot COOH + 2H_2O \rightarrow 2CO_2 + 4H_2$$

These photodecompositions take place fastest either with glutamate as the nitrogen source, or when the nitrogen source is exhausted. Since the energy yields are small, it is not likely that this is primarily a nutritive process. It might be the case that, as the hydrogenase enzyme is close to the site of nitrogen fixation, the evolution of hydrogen is a means of ensuring maximum activity for the fixation of nitrogen gas from the air. A strong reductant (ferredoxin rather than NADH) is required for nitrogen fixation, and since ferredoxin has a lower potential than the standard hydrogen electrode at pH 7, activation of the nitrogen fixing process would be expected to result in hydrogen gas liberation by the hydrogenase. Both the 'nitrogenase' and 'hydrogenase' systems are inhibited by ammonium ions, which is further evidence that the systems are related.

One must conclude this chapter by noting that just as the photosynthetic bacteria have been shown to be more and more diverse in their metabolic pathways, so one might expect the green plant to diversify the range of its products; some indication of this will be discussed in Chapter 10. In biochemistry, however, the bacteria generally show themselves capable of more diverse behaviour and of a greater degree of adaptation to nutritional circumstances than animals and plants. It is nevertheless surprising that the photosynthetic bacteria remain anaerobes, for this limits them to a life in small, isolated habitats, usually on mud under stagnant water, in contrast to the (aerobic) algae which have freedom to live in the open sea.

Reference

Evans, M. C. W., B. B. Buchanan and D. I. Arnon (1966). *Proc. Nat. Acad. Sci.*, **55**, 928.

Problems

Numerical work, as far as possible based on actual experimental results, covering and extending the material of chapters 1–5. The necessary physical constants, etc., will be found in the Appendix. Answers are on p. 185.

Problems

1. A leaf of *Beta vulgaris* may contain 18 μg chlorophyll per square cm of surface area. If it is illuminated by full sunlight (say 340 000 erg cm^{-2} sec^{-1} visible light energy, of mean wavelength 560 nm) and absorbs 50% of the light, how often does any single molecule of chlorophyll absorb a photon? If the average lifetime of fluorescence of chlorophyll *a in vivo* is 0·7 nsec, what fraction of chlorophyll molecules are excited at any given time?
   Molecular weight of chlorophyll $a = 892$

2. Leaves of *Nicotiana tabacum* L. were subjected to flashes of intense light and very short duration; the fixation of CO_2 was measured as a function of the frequency of flashing (following the principle of the experiments of Emerson and Arnold, see p. 36) and the maximum uptake per flash estimated. The results appeared to be widely scattered but on inspection, and with a sufficient number of repetitions, it was seen that the values were grouped closely round five mean values (see Table, col. I). What could such a distribution mean?

Table

	I Normal variety			II Mutant strain		
Group	mean value*	S.D.	% of total	mean value	S.D.	% of total
a	4·13	± 0·60	9	3·74 ± 0·58		16
b	1·78	± 0·07	20	1·86 ± 0·06		46
c	0·99	± 0·03	29	0·94 ± 0·05		16
d	0·44	± 0·01	31	0·42 ± 0·03		19
e	0·22	± 0·01	11	0·21 ± 0·02		3

* mean value of μmoles CO_2 fixed per g chlorophyll per flash.
S.D. indicates the standard deviation of each result.

The molecular weight of chlorophyll *a* is 892.

Column II of the table shows the distribution obtained with the *aurea* mutant of the tobacco variety above. What difference is observed? Comparable effects were obtained by varying the age and physiological state of the material used. What conclusions can be drawn from this work?

3. The energy of activation of a chemical reaction can be estimated by measuring the velocity constant at different temperatures, and applying the equation

$$k = Ae^{(-E_a/RT)}$$

where k is the velocity constant at temperature T (absolute), R is the gas constant and A is the Arrhenius constant, specific for the reaction. The delayed light emission (see p. 38) of an algal suspension was measured at temperatures between 10° and 40°, and the relative intensities are given below. If the intensity is proportional to the velocity constant of some chemical or electrochemical process, calculate the energy of its activation. Secondly, given that the emission originates from an energy store in System II, calculate the energy per electron in the store, if the wavelength of the emitted light is centred around 690 nm.

Temp (°C)	10°	20°	30°	40°
Relative intensity	1·0	4·1	15·1	51·3

4. Distinguish between the free energy change (ΔF) for a reaction, and the standard free energy change (ΔF_0).

When a suspension of chloroplasts was incubated in the light with ADP, ATP and phosphate in a KCl medium containing sufficient ferricyanide and 10 mM Mg^{++} at pH 7·5 and 30°C, the steady-state concentrations of the following reactants after 20 min were: ADP, $10\,\mu M$; ATP, 1·7 mM, and phosphate, 2·5 mM.

(a) What is the free energy change under these steady-state conditions for the synthesis of ATP:

$$ADP + P_i \dashrightarrow ATP$$

$$\Delta F_0 = +9\cdot6 \text{ kcal mole}^{-1} \text{ at pH } 7\cdot5, 30°C, 10 \text{ mM } Mg^{++}.$$

(b) If the synthesis of ATP was coupled to electron flow, what would you expect to be the redox potential span across a coupling site?

(c) Where, in non-cyclic electron flow from water to ferricyanide ($E_0' = +0\cdot42$ at pH 7·5) is there sufficient energy for the synthesis of ATP at the above concentrations of the reactants?

5. Chromatophores prepared from the photosynthetic bacterium *Rhodopseudomonas capsulata* showed on illumination a red-shift in the absorption spectrum of their carotenoid pigments, measured as a change in extinction at 530 nm. The shift did not take place in the presence of uncoupling agents. What could be inferred from this?

Chromatophores were suspended in the dark in a potassium-free salt medium (100 mM choline chloride) at 25° in the presence of valinomycin. Valinomycin renders chromatophore membranes permeable to the K^+ ion, while they remain impermeable to all the other ions present. When K^+ was added to the medium (as KCl) a carotenoid shift was seen which was similar to that previously observed on illumination. The table shows the extent of the shift with various concentrations of KCl, and under illumination. What information about the cause of the shift could you derive from this experiment? (Note that the entry of K^+ into the chromatophores would be opposed by a membrane potential ψ according to the equation

$$\psi = \frac{RT}{F} \ln ([K^+]_{outside} / [K^+]_{inside})$$

where the symbols have their usual meaning, ψ is given in volts and $[K^+]_{inside}$ remains virtually constant.

Table

Conditions	Change in extinction at 530 nm
No additions, steady state of illumination	+0·3
KCl added in dark, to final concentration shown, in the presence of valinomycin:	
0·1	+0·02
1·0	+0·08
2·0	+0·10
5·0	+0·12
10	+0·15
50	+0·18

6. The radioactivity of ^{14}C CO_2 fixed by the enzyme RuDP-carboxylase appears in the carboxyl-carbon of PGA. By means of a diagram such as Figure 5.1, show the distribution of label among the carbon atoms of the intermediates of the reductive pentose cycle after 1, 2 and 3 turns.

   In an actual experiment the following activities were recorded for the carbon atoms of hexose (a) at 5 sec exposure, and (b) at 30 sec.

	C1	C2	C3	C4	C5	C6
(a)	0·05	0·054	0·73	1·0	0·008	0·008
(b)	0·2	0·17	0·86	1·0	0·14	0·18

Is this result in accord with your prediction above? Comment.

7. It has been calculated that the chlorophyll content of a single chloroplast is of the order of $2·5 \times 10^{-12}$ gm. Assuming that the chlorophyll content of a spinach leaf is 0·1% of the fresh weight, calculate the number of chloroplasts in a spinach leaf weighing 5 gm (fresh weight).

8. Assuming that you can exist on a diet providing 2400 Kcal/day and supposing that this requirement could all be met by sucrose (at 4 Kcal/gm), how many 5 gm spinach leaves must work 10 hours a day (fixing CO_2 at 100 μmoles/mg chlorophyll/hour) to keep you alive?

9. A chloroplast differs in size from the wavelengths it absorbs by how many orders of magnitude?

Topics for discussion or requiring extended treatment.

Discuss the use of the term "efficiency" in photosynthesis. How can the term be applied (1) to the essentially irreversible reduction of NADP by illuminated chloroplasts and (2) to the conservation of energy in plants?

Comment on the usefulness of attempts to define the 'photosynthetic unit'.

What are the energy requirements for CO_2 fixation? Indicate how these requirements may be provided by green plants.

Describe the measurement of quantum efficiency in whole cells. Of what value are such measurements?

What is the evidence to support the concept that photosynthesis involves the co-operation of two distinct light reactions? (See Chapter 6.)

Comment on the relationships between the formation of proton and ion gradients and the synthesis of ATP in chloroplasts and chromatophores. (See Chapter 9.)

Compare the overall electron transport pathways in green plants and photosynthetic bacteria.

What is the evidence that chloroplasts are the site of photosynthesis in plants?

A series of compounds isolated from chloroplasts may be listed in order of decreasing redox potential. Of what value is this list in investigations of energy conservation in plants?

Do the pathways of CO_2 fixation in green plants and photosynthetic bacteria differ in any essential way?

Certain marine molluscs contrive to transfer live chloroplasts from their food plant into their own cells. Would you expect the slugs to obtain a major contribution to their energy from light? What factors need to be taken into account? See Taylor, D. L. (1971). *Comp. Biochem. Physiol.*, **38A**, 33.3

Part 2

CHAPTER 6

Evidence for two light-reactions in photosynthesis in green plants

Emerson and Arnold, in 1932, conceived of the photosynthetic unit as a group of some 2500 chlorophyll molecules coupled to a chemical reaction centre where carbon dioxide was reduced to carbohydrate and oxygen was evolved. Now the position is seen to be more complicated: the reduction of carbon dioxide is seen to depend on an electron transport process, and it is an electron movement itself which is the primary effect of the absorption of light energy. This light-driven movement takes place at two centres possessing some 600 chlorophyll molecules together.

Evidence for these newer ideas accumulated during the decade from 1950, much of it obtained by Emerson himself. The two-light-reaction theory appeared first in 1960 (R. Hill and F. Bendall) and was based on theoretical considerations of the role of the chloroplast cytochromes. In 1961 Duysens and coworkers presented what was to become, essentially, the modern theory, on the evidence of their spectrographic observations; Duysens also allotted the terms photoreaction I, photoreaction II to the NADP-reducing and oxygen-evolving processes respectively.

During the next two or three years the new hypothesis was found to account for and predict many experimental observations. At the same time the 'zig-zag' pathway of electron transport containing the two light reactions in series was found to be the most useful embodiment of the hypothesis. The reader should however bear in mind that while one-light-reaction theories are no longer tenable, leaving the two-reaction principle as the simplest available, it cannot be said with so much certainty that the zig-zag arrangement is the best form of it (see sections 8.22 and 8.3). However, no alternative has had a success approaching that of the zig-zag formulation, just as no compelling evidence has been adduced for the existence of three light reactions or more.

In this chapter we shall review the various lines of evidence that today form the basis for our belief that two light reactions are indeed involved in photosynthesis in green plants. We can distinguish four separate topics: first, the

93

observations of differing effects when illumination was varied in wavelength; second, the argument of Hill and Bendall concerning the role of cytochromes and other redox carriers; third, the preparation of sub-lamellar particles by various means from chloroplast thylakoids, which appear to contain separated photoreaction systems; and, lastly, a theoretical argument based on the light energy required to reduce carbon dioxide (or to split water).

6.1 Experiments involving light of varied wavelength

In spite of the undoubted fact that chlorophyll *a* is the key pigment in both photoreactions of green plant photosynthesis, the proportion of light absorbed by each of them does not remain constant while the wavelength is varied. Had this single principle not been true, we might well have been unable to discover the existence of the two light reactions. The mechanism of the effect may lie to some extent in a different degree of association of the accessory pigments, carotenoids, and phycobilins or the chlorophylls *b*, *c* etc., with each of the two types of photosynthetic unit. Probably the more important point however is that chlorophyll *a* as it occurs *in vivo* is heterogeneous, in that various differences of environment or aggregation cause the absorption spectrum to be broadened by the addition of displaced components known as C_{670}, C_{680}, C_{690} and C_{700}. The extent to which these make up the two photosynthetic units may vary from species to species, and during the time course of illumination, as will be discussed in the next chapter. Nevertheless the chlorophyll form that absorbs furthest to the red, C_{700}, can be shown to be almost entirely part of photosystem I, while accessory pigments such as chlorophyll *b* in higher plants and phycobilins in the blue green and red algae appear to belong more to system II than to system I. It is however true that for the most part the experiments in this section demonstrate that two light reactions are a minimum condition of photosynthesis, but not that two light reactions are sufficient.

6.11 The 'red drop', 'enhancement' and chromatic transients

Early observations showed that the action spectrum (defined in section 2.6) for photosynthesis fell sharply at wavelengths above 680 nm, more steeply than the absorption of the pigments, which continued above 700 nm. Even allowing for the decreased accuracy of absorption and rate measurements at long wavelengths, it was clear that quanta absorbed by chlorophyll above 680 nm were less effective than those of shorter wavelength. This drop in quantum efficiency was termed the 'red drop', and an example is given in Figure 6.1.

In Figure 6.2, taken from Avron and Ben-Hayyim (1969), isolated chloroplasts were examined and the quantum requirement measured for

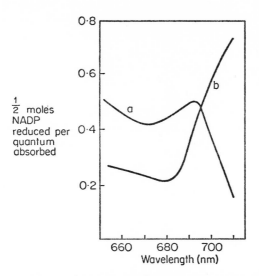

Figure 6.1. The quantum yield for the reduction of NADP by chloroplasts with electrons supplied from (a) water and (b) an artificial donor system. Note the 'red drop' in the former case. From Hoch, G. and I. Martin (1963). *Arch Biochem. Biophys.*, **102**, 430, with permission. Copyright held by Academic Press

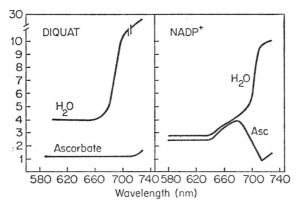

Figure 6.2. The quantum requirements for photosynthetic electron transport from water *or* ascorbate to NADP *or* diquat (a viologen compound). Note the sharp rise in the requirement when water is the donor, also the apparent change in the relative distribution of energy under the two sets of conditions. From Avron, M. and G. Ben-Hayyim (1969). In H. Metzner (Ed.), *Progress in Photosynthesis Research*, vol. 3, Institut fur Chemische Pflanzenphysiologie, Tubingen, p. 1185, with permission

electron transport from water or an artificial donor such as ascorbate to NADP, or an artificial acceptor such as diquat. At wavelengths above 670 nm the quantum requirement for electron transport from water rises very sharply, apparently without limit, while with ascorbate as electron donor the requirement falls to (or remains at) unity. These results, which had been known for several years, indicate that light of the longer wavelength is absorbed by pigments which are not able to apply the energy to the water–oxygen reaction, but which are coupled with high efficiency to the reduction of NADP or diquat.

Avron's purpose in the paper referred to was to show that the quantum requirements of various electron transport reactions depended closely on the conditions of the experiment, implying that the two kinds of photosynthetic unit were not to be thought of as structurally distinct and functionally isolated, but 'plastic' in that the distribution of quanta was controlled (at wavelengths less than 670–680 nm) in some manner, tending to make the most efficient use of the available light. We shall return later to the concept of the plastic photosynthetic unit.

Emerson showed with algae that the quantum efficiency of light in the far-red region was enhanced by a weak background illumination with shorter wavelength light. Enhancement was said to occur when the rate with the two beams of light together was greater than the sum of the rates with either beam alone. The effect, also known as the Second Emerson Effect (the first was concerned with the carbon dioxide 'gush' occurring with intact algae at the onset of illumination) has been well documented for algal cells, and for chloroplasts with intact envelopes, when either oxygen evolution or carbon dioxide fixation was measured. Some interesting results have been obtained by the study of *enhancement spectra:* these display the degree of enhancement of a fixed wavelength background illumination by a variable-wavelength beam. The resulting spectra show peaks which can be reasonably well identified with the pigments of the cells. Figure 6.3 (Fork, 1963) shows enhancement spectra for several species. It can be seen that chlorophylls *b* and *c* enhance the absorption by chlorophyll *a*, and also that the blue-absorbing peak of chlorophyll *a* at 436 nm follows the far-red absorbing band at 690–700 nm.

Enhancement is usually measured by selecting two wavelengths, λ_1 and λ_2 and obtaining the rates of reaction with each beam singly (R_{λ_1}, R_{λ_2}), and with both together ($R_{\lambda_1 + \lambda_2}$). The value for enhancement may be given either as a percentage (formula 1), where the condition of no-enhancement is a zero result, or as a simple ratio (formula 2), where the same condition is indicated by the value 1·0:

$$\text{Enhancement} = \frac{(R_{\lambda_1 + \lambda_2} - R_{\lambda_1} - R_{\lambda_2}) \times 100}{R_{\lambda_1} + R_{\lambda_2}} \tag{6.1}$$

$$\text{Enhancement} = \frac{R_{\lambda_1 + \lambda_2}}{R_{\lambda_1} + R_{\lambda_2}} \tag{6.2}$$

The values obtained are small, being under most conditions less than 2 or 3 (formula 2). This is an indication that quanta of wavelength less than say 670 nm are efficiently distributed between the two photoreactions by a variable spillover mechanism.

The third group of experiments in this section is the measurement of *chromatic transients;* these are sudden and shortlived jumps or drops in the rate of, say, oxygen evolution when the wavelength (but not the effective intensity) of the illumination is suddenly changed. This is the Blinks effect (L. R. Blinks, 1957) and was pursued intensively by French and Myers. This effect can be considered as an Emerson-enhancement experiment in which the two wavelengths are applied successively rather than simultaneously. The dark interval between the two light beams may be extended, when

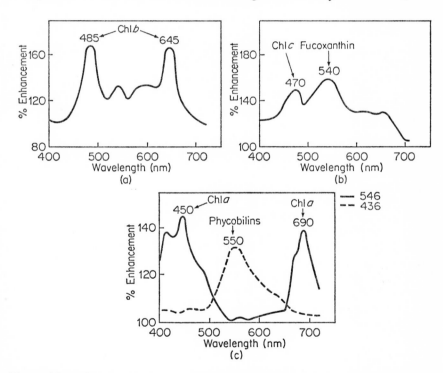

Figure 6.3. Enhancement in green, brown and red algae. (a) *Ulva*, with background light of wavelength longer than 680 nm. Note the prominence of the bands of chlorophyll *b*. (b) *Endarachne binghamiae*, with a background as in (a). Note the contribution of Chl*c* and fucoxanthin. (c) *Porphyra perforata* with two conditions of background. From Fork, D. C. (1963). In *Photosynthetic Mechanisms of Green Plants* Publ. No. 1145 NAS-NRC Washington, D.C. p. 352, with permission

the magnitude of the transient is found to diminish, suggesting that a chemical intermediate accumulates by the operation of one light-reaction, and is removed by the other. Quantitatively the experiment is hard to evaluate, since many effects seem to be involved. Thus in whole algal cells there is a light-stimulated uptake of oxygen (photorespiration) and under some conditions the production of oxygen is preceded by a temporary uptake.

6.12   Antagonistic effects of light beams of different wavelengths

The observations of the previous section showed that two light beams of differing wavelengths cooperated in the process of photosynthesis. The present section is concerned with an apparent antagonism. Thus illumination by far-red light (that is light of more than say 700 nm in wavelength) causes the oxidation of cytochrome f, plastocyanin, P700 and plastoquinone (see for example Witt, 1968).* In addition, far-red light acts to diminish the degree of fluorescence obtained at room temperature from flashes of light at 620–680 nm. The level of fluorescence emission from system II pigments is thought to be under the control of a 'quencher', Q, which is inactivated and activated reversibly; in the absence of far-red light, light of 620–680 nm produces immediate fluorescence at a level indicative of the state of Q at the time, and the emission increases in intensity up to a maximum as Q is inactivated. (Further, slower changes occur after that, which Duysens considers to indicate a rearrangement of chlorophyll molecules from system II units to system I.)

These results are compelling evidence that there are at least two separate kinds of photoreaction centres in the chloroplasts of higher plants, and that there are significant differences in the absorption spectra of the pigments attached to them. The zig-zag scheme provides a convenient rationale of the observed antagonistic effects of the two wavelengths: if the two light reactions and the redox carriers are in a linear series, and if one of the light reactions becomes rate limiting, the carriers on the source-side of it tend to become more reduced, and those on the acceptor-side more oxidized. Changing the wavelength of the illumination may cause the other light reaction to become rate-limiting instead, and the state of the carriers will change accordingly.

6.2   The need to accommodate redox materials of intermediate potential

The chloroplast cytochromes, f, b_6 (both 559 and 563 nm components) and the high potential cytochrome-559 all have standard redox potentials at pH 7 in the range 0–0.4 V. Plastocyanin, P700 and the plastoquinones also have

---

* The observations on cytochrome f were the basis of the distinction between systems I and II originally made by Duysens and coworkers (1961).

standard potentials in or close to that bracket. Probably the first clear suggestion of the zig-zag scheme was that given by Hill and Bendall in 1961, on the basis of the argument that any one-light-reaction theory accounting for electron transport from water (+0·82 V at pH 7) to NADP (−0·32 V) would only use electron carriers with potentials outside that range, as

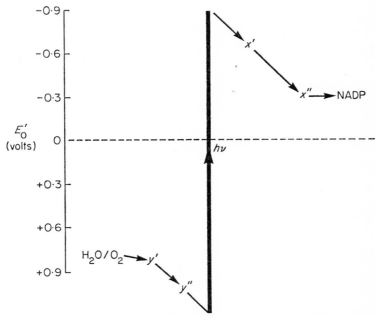

Figure 6.4. To show that redox materials involved in a one-light-reaction electron transport process from water to NADP must have E_0' values outside the range
+0·82 V−−0·32 V

indicated in Figure 6.4. (Remember that in these diagrams spontaneous electron flow is represented by a downward arrow, upward arrows indicating a light-driven step.) Hill and Bendall showed that cytochromes b_6 (0·0 V) and f (0·37 V) could be accommodated by a two-light-reaction scheme (Figure 6.5). There were at that time some indications that f was oxidized and reduced by illumination under various conditions, and these have since been well established. A further reason for involving cytochromes was that the phenomenon of photophosphorylation had been recorded by Arnon's group, and Hill and Bendall were conscious of the possible analogy between that and oxidative phosphorylation. An intimate account of the development of these concepts has been given by Hill (1964), and also by Arnon, (1968).

6.21 The existence of two series of electron acceptors

Kok (1966) has given an account of various electron acceptors in photo-synthetic electron transport (Hill oxidants). Some of these can be arranged into two series. One series, the viologen dyes, have low potentials ($-0\cdot318$ V to $-0\cdot740$ V) and are reduced by chloroplasts according to Table 6.1, where it can be seen that there is some limiting potential at approximately $-0\cdot65$ V. Zweig and Avron (1965) have also reported a similar limit, at $-0\cdot5$ V, and

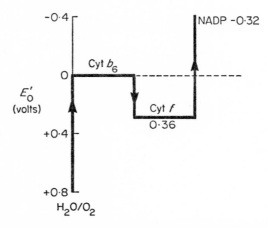

Figure 6.5. The first formulation of the 'zig-zag scheme'. From Hill, R. and F. Bendall (1960), *Nature*, **186**, 136 with permission

Black (1966) at $-0\cdot52$ V. A second series, of substituted quinones, was applied to a mutant alga, *Scenedesmus* No. 8, which was unable to carry out the far-red photoreactions of photosynthesis. The quinones were reduced by the Hill reaction, down to a limit, this time at a potential of $+0\cdot17$ V. A more sensitive method was found, in which Kok and his colleagues measured the intensity of fluorescence, which comes mainly from system II at room temperature, and which is governed by the redox-state of the quencher, Q. By adjusting the concentrations of oxidized and reduced naphthoquinones, making redox buffers at various potentials in which chloroplasts were suspended in the dark, the initial level of fluorescence was found to vary, indicating that Q responded to the redox potential of the solution. Since there should have been equilibrium in the dark period, the external potential at the point of 50% quenching of fluorescence indicated the redox potential of Q: a value of $+0\cdot18$ V was obtained. Cramer and Butler (1969) have extended this approach and have produced values for the redox potential of Q of between -20 mV and -35 mV (see Figure 8.4). It was necessary to use the mutant since otherwise the low-potential reductants produced by photoreaction I would have interfered with the higher potential product of reaction II. It can be concluded that the electron acceptors are reacting at two distinct sites, which have widely differing redox potentials. The more negative of these is low enough to account for the reduction of NADP, but the more positive potential has no

Table 6.1. Two series of Hill reagents.

Series I—viologens		Series II—quinones		
E_0 (mV)	% reduced		E_0 (mV) pH 6·5	O_2 evolution
-318	100	$K_3Fe(CN)_6$	$+420$*	$+$
-440	100	p-Benzoquinone	$+330$*	$+$
-548	100	Methyl-p-benzoquinone	$+280$	$+$
-670	26	2,5-Dimethyl-p-benzoquinone	$+220$	$+$
-740	4	2,6-Dichloroindophenol	$+220$	$+$
		1,2-Naphthoquinone†	$+190$*	$+$
		1,2-Naphthoquinone + sulphate†	$+160$*	*weak*
		1,4-Naphthoquinone	$+90$	$-$
		2-Methyl-1,4-Naphthoquinone (menadione)	$+20$	$-$
		2,5-Dibydroxybenzoquinone	-80	$-$
		2,Hydroxy-1,4-Naphthoquinone	-90	$-$
		Sodium anthraquinone sulphonate	-180	$-$
		FMN	-180	$-$
		Methyl viologen	-440	$-$

* At pH 6·6
† Data from Kok (1966) and Kok and Datko (1965)

explanation on a one-light-reaction theory. If on the other hand the two-light-reaction hypothesis is accepted, these measured potentials give an indication of the potentials of the primary reductants produced by the two photoreactions: $-0·5$ V (system I) and $-0·03$ V (system II).

6.3 Fractionation of the thylakoid membrane

Treatment of the chloroplast with surface active agents often results in its disruption into small sub-lamellar particles. Thus Boardman and Anderson found that *digitonin* allowed the separation by differential centrifugation of two fractions, one containing small particles, with a higher percentage of chlorophyll *a*, which were more active in the reduction of NADP by ascorbate, and a residue of much larger particles, having a lower ratio of chlorophyll *a*/*b* than in the original chloroplasts and which was the more active in the Hill reaction (the production of oxygen with ferricyanide as electron acceptor). Since the total recovery of the two activities was good, it was claimed that the digitonin had physically divided the thylakoid into constituent parts, and that each part clearly carried out a different photo-reaction. This work has been developed by several workers and will be reviewed in the next chapter. For our purposes here, it can be said that particles which carry out the ascorbate–NADP photoreaction (I) have been prepared by several methods and refined to small size, containing little more than protein, chlorophyll *a* and P700. However for the full proof the same should be done with the Hill reaction, and this appears to be impossible, because once the thylakoid is broken down to small particles, apparently by any means, the Hill reaction is lost. There is some reason to believe that the Hill reaction requires a vesicular structure and the smallest particles prepared that have been Hill-active were vesicles of some 50 nm diameter. Smaller particles than that are easy to obtain, but are inactive.

6.4 Energetics and the quantum yield

From the second law of thermodynamics, any apparatus which irreversibly converts one form of energy into another must show a degree of inefficiency, that is, the energy stored must be less than the energy supplied. The difference appears as liberated heat (as in the effect of friction) and as an increase in disorder. A photosynthetic theory must allow for this law: the total energy of the quanta of light absorbed must be more than the total energy change represented by the chemical reaction carried out.

Light of 680 nm has an energy of approximately 41 kcal per einstein. Since the standard free energy change in the reduction of carbon dioxide to carbohydrate, with the production of oxygen from water is 120 kcal mole^{-1}, 3 quanta per molecule of carbon dioxide would achieve a balance, and 4 quanta would give a figure for the efficiency of the system of $120/164 = 73\%$, which is high (but not inconceivable). Since however we know that the complete chemical process involves many steps, we can refine our estimate by considering the energy trapped by the primary photoprocess itself. For this purpose the energy of the 680 nm quantum is better expressed as approximately 1·8 electron volt (eV) per photon. The difference in the redox potentials of the primary electron donor(s) and acceptor(s) should not exceed 1·8 V for a one-quantum process. There is a reasonable estimate of the potential achieved at the reducing side, based on observations of the degree to which chloroplasts will reduce artificial (viologen) electron acceptors: the value is approximately -0.52 V. The oxidizing potential achieved is difficult to estimate since the primary donor is not known; but it must be at least (say) 100 mV more positive than the standard potential of the oxygen–water couple ($+0.82$ at pH 7) since there seems to be no tendency for the evolution of oxygen to reverse. The minimum difference between the two potentials is 1·44 V to which we must add the energy stored as ATP. The standard free energy of ATP formation from ADP and P_i is 7·4 kcal mole^{-1}, or 0·3 eV per molecule for a one-electron process. We need to increase this figure to allow for the fact that the process of photophosphorylation is able to achieve a high proportion of ATP to ADP, and this may mean that the actual, as opposed to the standard, energy is more than 0·6 eV per molecule (or 0·03 eV for each of a pair of electrons per molecule). Furthermore, there is some reason to believe that there are 2 molecules of ATP formed for the passage of each pair of electrons, instead of one as previously thought. If this is the case, the allowance for photophosphorylation is at least 0·6 V, to be added to the redox potential difference of at least 1·44 V. The total is more than 2·0 V, which is impossible if only one quantum is required for the passage of one electron.

Even if we suppose that phosphorylation is less complete in the system involving artificial electron acceptors (there is little reason why this should be so) or take the lower value for the ATP/2e$^-$ ratio, we still obtain a total energy storage approaching the 1·8 eV supplied, sufficiently close to throw doubt on a one-quantum mechanism.

Another way of looking at the matter is to consider the quantum yield measurements for photosynthesis, in algal cells. Many workers have consistently measured maximum yields of about 0·1, sometimes as high as 0·11, molecules of oxygen evolved per photon absorbed. It is more satisfying to consider 0·11 as an experimental approximation to 0·125 indicating that eight quanta are required for the four electron process, rather than to suppose that the efficiency of the pigment system itself is less than 50%.*

If it is accepted on the evidence reviewed in this chapter that there are two light reactions in photosynthesis, then the observations that follow can be applied to the discussion of the form the mechanism must take. We shall examine the structure of the thylakoid more closely, and then review electron transport and the evidence concerning the choice of the zig-zag or other models.

* Warburg has for many years claimed that yields of some 0·25 or higher can be obtained under the right conditions. If this is the case, the two light reaction theory is weakened severely; however most workers consider these high values to be experimentally unsound. A paper by Emerson (1947) indicates the ground on which this criticism is usually based. Warburg has on the strength of his measurements put forward a one light reaction theory which will be reviewed in Chapter 8.

References

Arnon, D. I. (1968). In T. P. Singer (Ed.), *Biological Oxidations*, Interscience, New York, p. 123.

Avron, M. and G. Ben-Hayyim (1969). In H. Metzner (Ed.), *Progress in Photosynthesis Research*, Vol. 3, Institut für Chemische Pflanzenphysiologie, Tubingen, p. 1185.

Black, C. C. (1966). *Biochem. Biophys. Acta*, **120**, 332.

Blinks, L. R. (1957). In H. Gaffron and others (Eds.), *Research in Photosynthesis*, Interscience, New York, p. 444.

Cramer, W. A. and W. L. Butler (1969). *Biochim. Biophys. Acta*, **172**, 503.

Duysens, L. N. M., J. Amesz and B. M. Kamp (1961). *Nature*, **190**, 510.

Emerson, R. and W. Arnold (1932). *J. Gen. Physiol.*, **15**, 391 and **16**, 191.

Emerson, R. and M. S. Nishimura (1949). In J. Franck and W. E. Loomis (Eds.), *Photosynthesis in Plants*, Iowa State College Press, Ames, Iowa, p. 219.

Fork, D. C. (1963). In *Photosynthetic Mechanisms of Green Plants*, Publ. No. 1145, NAS-NRC, Washington, D.C., p. 352.

Hill, R. (1965). In P. N. Campbell and G. D. Greville (Eds.), *Essays in Biochemistry*, Academic Press, London, p. 121.

Hill, R. and F. Bendall (1960). *Nature*, **186**, 136.

Hoch, G. and I. Martin (1963). *Arch. Biochem. Biophys.*, **102**, 430.

Kok, B. and E. A. Datko (1965). *Plant Physiol.*, **40**, 1171.

Kok, B. (1966). In J. B. Thomas and J. C. Goedheer (Eds.), *Currents in Photosynthesis*, Donker, Rotterdam, p. 383.

Zweig, G. and M. Avron (1965). *Biochem. Biophys. Res. Commun.*, **19**, 397.

CHAPTER 7

The structure of the thylakoid membrane

The thylakoids of the chloroplast present a most striking appearance in electron micrographs, of closely oppressed pairs of membranes, often oriented together and stacked; the membranes are thicker, and stain more densely with most electron-dense stains than membranes from other organelles and tissues. Membranes currently form a very active topic of biochemical research, for three general reasons: they constitute a stable, solid phase, two-dimensional system in the liquid-phase, three-dimensional medium of the cell, and the materials of which they are made have interesting physical and chemical properties; secondly, since a membrane forms a barrier which resists the passage of many solutes and supports electric potentials, we may look for some special organization of the protein and lipid components; thirdly, virtually all subcellular membranes have a degree of catalytic activity, from the facilitated-diffusion systems involving 'permease' enzymes, and the existence of pores for the entry of specific materials, to active transport systems that perform osmotic work at the expense of ATP, and electron-transport systems that can conserve the energy of redox reactions.

The structure of the thylakoid membrane attracts interest not only in connection with the above, but also by virtue of the heterogeneity implied by the two light reactions of photosynthesis. We have seen that there are units of some 600 chlorophyll molecules partly or completely divided between the two reaction centres, and we might expect to find particles, one or perhaps two per 600 chlorophylls. Again, we ascribed differences in the reaction spectra of the two photoreactions to differences in the pigment types associated with each centre; if the pigments are functionally distinct we might expect to find a structural separation. Such a particulate understructure might be found either by fragmenting the thylakoid, or by using the electron microscope; both methods have yielded interesting results.

Membranes in cells, such as the cell envelope, the endoplasmic reticulum and the other and inner membranes of mitochondria and nuclei may have an underlying similarity. Thus possession of cytochromes has been demonstrated not only in mitochondrial inner membranes, but also in nuclear membranes and the endoplasmic reticulum, so that one might wonder if traces of electron transport carriers might be found in the cell membrane and in the chloroplast outer envelope. Although the inner membranes that form the cristae of the mitochondrion and the thylakoid of the chloroplast are thicker than the others, there is evidence that these membranes are formed by proliferation of the outer membrane of the organelle, the thickening occurring subsequently. This would imply a fundamental homology in the structures of thick and thin membranes. There are also observations that suggest that mitochondria and chloroplasts develop as outgrowths of the nuclear membranes, although discussion of this point is beyond the scope of this text.

7.1 Structural components

It will be obvious at the outset that the label 'structural component' tends to be applied to those materials for which no other function has been found. At the same time the structural properties of catalytically-active components are often overlooked. Table 7.1 (from Lichtenthaler and Park, 1963) gives a list of the principal components of thylakoid fragments, and their abundance (plastocyanin may well have been partly lost in the preparative procedure), from which it can be seen that the pigments occur in sufficient quantity to make a considerable contribution to the properties of the water-insoluble part of the matrix. Proteins such as the apoprotein of cytochrome f (molecular weight 62 000 per haem group) could be playing an additional, structural, role. However there is little doubt that a major structural feature is the '*structural protein*' which accounts for some 20–30% of the dry weight of the thylakoid. Criddle (see for example Criddle, 1965) prepared protein of this kind first from mitochondria, and then from chloroplasts, demonstrating the similarity. The method Criddle used was to remove the pigment from the thylakoids with acetone, and extract the dried powder with cholic and deoxycholic acids, finally fractionating the product with ammonium sulphate and sodium dodecyl sulphate (SDS). 'Structural protein' has also been obtained from the erythrocyte 'ghost', although this last differs from the others in that it is soluble outside much narrower pH limits. At moderate pH values the chloroplast protein is only soluble in detergents such as cholic acid, or sodium dodecyl sulphate. Structural protein of both chloroplast and mitochondria is characterized by a molecular weight in the region of 23 000, and a high proportion of non-polar amino acids. Criddle showed that it was able to bind many porphyrin-type compounds in stoichiometric proportions: chlorophyll, myoglobin and cytochromes. The detergent–chlorophyll–protein particles prepared by Thornber

Table 7.1. Representative distribution of substances in spinach chloroplast lamellae on basis of minimum molecular weight of 960 000 per mole of manganese (From H. K. Lichtenthaler and R. B. Park (1963), *Nature*, **198**, 1070, with permission).

Lipid (composition, moles/mole Mn)			
115 chlorophylls		103 200	
80 chl. *a*	71 500		
35 chl. *b*	31 700		
24 carotenoids		13 700	
7 β-carotene	3 800		
11 lutein	6 300		
3 violaxanthin	1 800		
3 neoxanthin	1 800		
23 quinone compounds		15 900	
8 plastoquinone *A*	6 000		
4 plastoquinone *B*	4 500		
2 plastoquinone C^{21}	1 500		
4,5 α-tocopherol	1 900		
2 α-tocopherylquinone	1 000		
2 vitamin K_1	1 000		
58 phospholipids		45 400	
(phosphatidylglycerols)			
72 digalactosyldiglyceride		67 000	
173 monogalactosyldiglyceride		134 000	
24 sulpholipid		20 500	
? sterols		7 500	
Unidentified lipids		87 800	
			495 000
Protein			
4690 nitrogen atoms as protein		464 000	
1 manganese		55	
6 iron		336	
3 copper		159	
			465 000*
	Lipid + protein		960 000

* Rounded off to nearest thousand.

and coworkers (1967b) can be regarded as two structural proteins with different amino acid compositions. The Criddle preparation may be a mixture of these two proteins.

The thylakoid has its complement of lipids, which are characteristic. Table 7.1 contains a list of these. The unsaturated fatty acids are interesting both for their abundance (said to be a feature of oxygen-evolving photosynthetic systems), and from their resistance to oxidation by the high partial pressure of oxygen which must obtain during photosynthesis. The resistance is often lost when chloroplasts are isolated. Apart from noting the existence of the characteristic thylakoid lipids (of which some formulae are given in Figure 7.1), and commenting that they may have an insulating and cementing-role in the membrane, there is little that can be said concerning their arrangement. Weier and Benson (1966) have however prepared for discussion a model structure in which various types of lipid are allocated a position in the thylakoid.

It should be noticed that in grana the outer surfaces of thylakoids adhere together, and continue to do so if the granum is osmotically swollen; the swelling takes place inside the thylakoid. Therefore, the outside of the

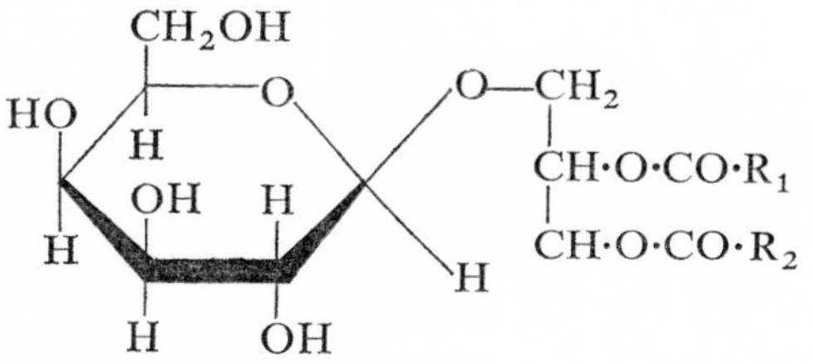

I Monogalactosyl lipid   40%

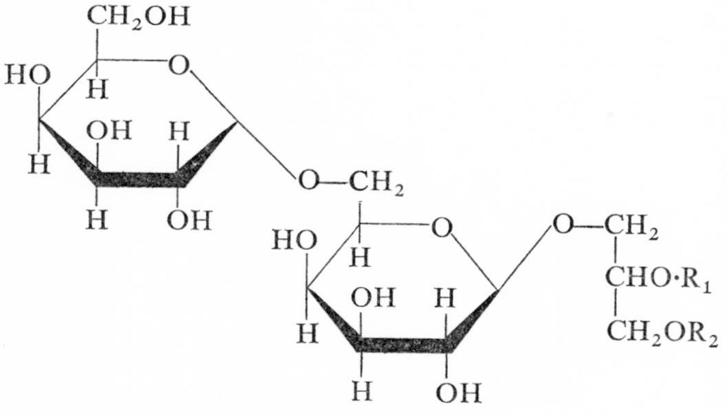

II Digalactosyl lipid

Figure 7.1. Formulae of some of the thylakoid lipids, from Weier, T. E. and A. A. Benson (1966). In T. W. Goodwin (Ed.) *Biochemistry of Chloroplasts*, Vol. 1, Academic Press, London, p. 91, with permission

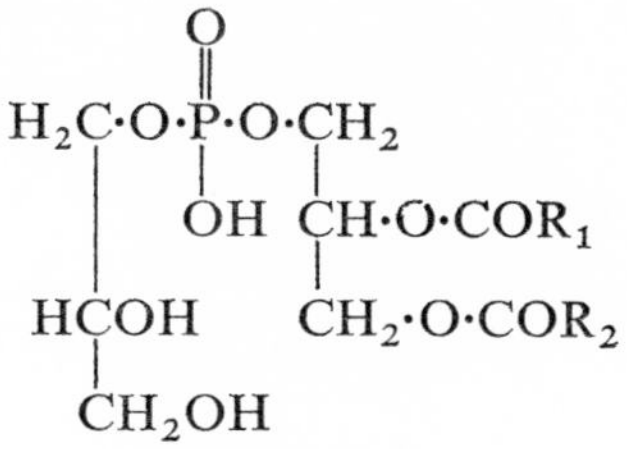

O
‖
$H_2C \cdot O \cdot P \cdot O \cdot CH_2$
| |
OH $CH \cdot O \cdot COR_1$
$HCOH$ $CH_2 \cdot O \cdot COR_2$
|
CH_2OH

III Phosphatidyl glycerol

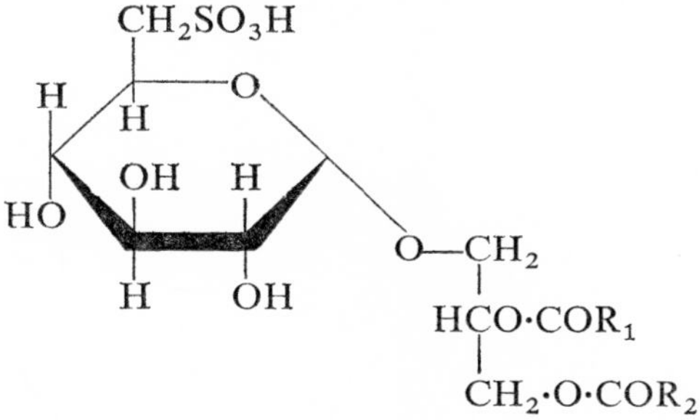

CH_2SO_3H

IV Sulphoquinovosyl diglyceride

R_1, R_2: major component—α linolenic acid

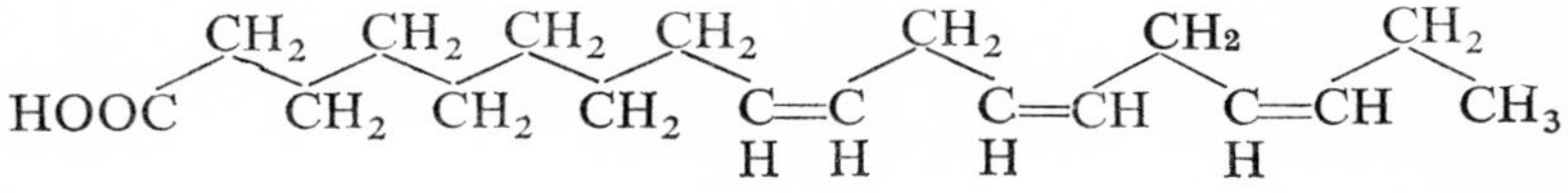

Minor component—*trans* $\triangle^3$ hexadecenoic acid

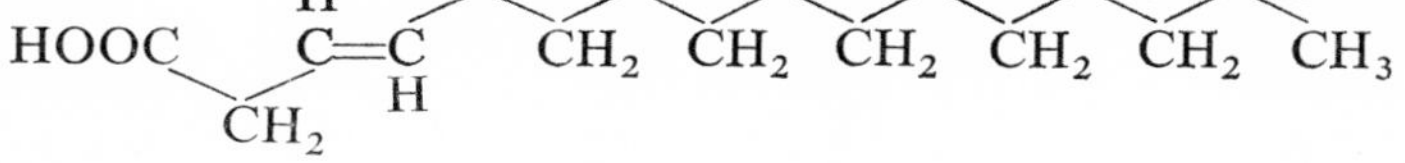

Figure 7.1. continued

109

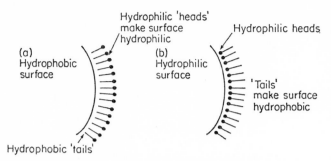

Figure 7.2. The action of surfactant lipids in coating either hydrophilic or hydrophobic surfaces

thylakoid is presumably hydrophobic and the inside hydrophilic. This does not however identify the site of lipid, since, as many of the lipids are hydrophilic at one end and hydrophobic at the other, the presence of lipid in a monolayer merely reverses the character of the previous surface (Figure 7.2).

7.2 Pigments

There is a degree of organization shown in the complement of pigments of photosynthetic structures. We are led to this conclusion by four lines of evidence: (i) the conclusions of Chapter 6, that there were two photoreactions, drawing their energy from separate groups of pigment molecules; (ii) resolution of sub-lamellar particles which differed in the proportions of chlorophyll a to b, and of carotene and xanthophyll pigments (see section 7.3); (iii) observations with polarized light that show part of the chlorophyll to be arranged in a different manner from the rest, and (iv) that the chlorophyll of the thylakoid can be resolved into discrete components identical chemically but having different absorption properties.

Olson (1963) has given a useful account of studies on the polarization of light absorbed and emitted by chloroplast material. In one approach, measurement of *dichroism*, use is made of the fact that a chromophore has a directional property: the electric vector of the light-waves must be in the direction of the electron transition. When pigment molecules are oriented together they will preferentially absorb plane-polarized light when the direction of polarization is parallel to this common direction. Rotation of the plane of polarization causes the light absorption to fluctuate; this effect is known as *dichroism*. Alternatively, if the material is illuminated with unpolarized light, it absorbs one plane of vibration preferentially, so that the light that passes through is partially polarized by subtraction. When algal cells were examined in this way, and the polarization determined by rotating a Nicol prism behind the microscope objective, it was found that dichroism was only observed at the edge of a chloroplast, that light with its plane of vibration tangential to that edge was preferentially absorbed, and that the effect disappeared when the chloroplasts were damaged in any way. Furthermore, using monochromatic light, it was found that the dichroic effect was at a maximum between 695 and 705 nm. (The wavelength of maximum absorption is usually 675–680 nm.) There are technical problems both with obtaining unpolarized monochromatic light, and in making the observations at a wavelength at which human vision is almost impossible. The conclusion drawn from this experiment is that the thylakoid membranes, which in intact cells are parallel with the surfce of the chloroplast, contain some 5% of the chlorophyll in a long-wavelength-absorbing form, and the molecules themselves are oriented in the plane of the thylakoid.

Bifluorescence is an analogous effect: an oriented pigment array emits radiation with the electric vector again parallel with the direction of the electron transition, so that the fluorescence is plane-polarized, and the apparent intensity of the fluorescence can be varied by rotation of an analyser through which it is observed. If the fluorescence is excited by mercury light (436 nm), which is convenient as transmitted light can be stopped by a filter, the longer wavelength (720 nm) components of the fluorescence are polarized, in the same direction as previously. If the mercury light is polarized, the direction of the plane has no effect on the fluorescence in intensity or in polarization. If however a ruby laser is used for the light-source (694·3 nm), polarized fluorescence is again observed at 720 nm (most of the fluorescence is at 685 nm), but now it is found that rotation of the laser beam (which is naturally plane-polarized) varies the intensity of the fluorescence. One explanation for the difference between the effects of mercury and ruby light is that with the former, energy reaches the oriented fraction of chlorophyll after many transfers from the original absorbing molecules, and during this time the polarization is lost. With the ruby, the wavelength (694·3 nm) stimulates the dichroic chlorophyll (695–705 nm) directly; the fluorescence emerges with its polarization unchanged. Excitation energy from the oriented chlorophyll cannot easily stray into the main unoriented pigment mass since the 10–20 nm wavelength difference imposes an energy barrier. These experiments also indicate that the oriented chlorophyll is not able to move around freely and thus change the direction of polarization of its fluorescence.

We have had occasion to distinguish part of the thylakoid chlorophyll by means of its absorption and fluorescence maxima. The idea that chlorophyll in the chloroplast was heterogeneous appears to have begun with Albers and Knorr, who in 1937 claimed that inspection of the absorption spectra of certain preparations revealed the presence of components with maxima at 674 and 683 nm, and possibly others at 668, 687 and 698 nm. Since the overall absorption peak, in the red, of chlorophyll *in vivo* is so much broader than in solution, even after allowing for the effects of light scattering, this was an attractive concept, and the work has since been confirmed many times. French and his coworkers have recently set up a computer routine for fitting idealized chlorophyll spectra to the actual spectra of algal cells. An example of the kind of result obtained is given in Figure 7.3. Previously the same group made considerable progress with a differentiating spectrophotometer; this instrument recorded, instead of the extinction, E, the derivative $dE/d\lambda$, against the wavelength. This has proved easier to interpret than the absorption curves themselves, at least without computer assistance. An example is given in Figure 7.4. Work of this kind has been carried out in several laboratories, and a consensus has grown up that there

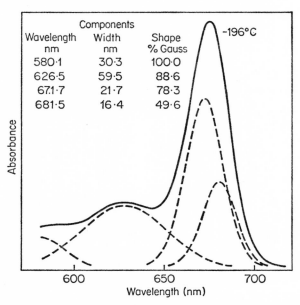

Figure 7.3. The absorption spectrum of a pale mutant of *Chlorella*, from which a computer program has extracted the 'components' as shown. From French, C. S. and L. Prager (1969). In H. Metzner (Ed.) *Progress in Photosynthesis Research*, Institut fur Chemische Pflanzenphysiologie, Tubingen, p. 558, Fig. 2, with permission

are at least three components, the first two being major, each constituting close to 50% of the total chlorophyll. The wavelength ranges of these two are generally agreed to be 670–673 nm and 680–683 nm; both the precise wavelength and quantity differ from species to species, and in different preparations. The third type absorbs variously in the range 695–705 nm or even beyond. Opinion is at present divided on whether this is a single class or not; we will argue here that it makes better sense if the 695 and 705 nm components are separate. The 695–705 nm form(s) constitute some 3–5% of the total chlorophyll. We should distinguish also the photoreactive pigment P700, which is reversibly bleached by light or by oxidation, and is believed to be chlorophyll at the active centre of system I. Kok, the discoverer of P700, estimated it to be some 0·3 % of the total chlorophyll; its contribution to absorption or fluorescence phenomena must be negligible. The same applies to P690, which is a similar entity claimed by Witt's group to be the active centre for system II, on the basis of fast-reaction spectroscopy.

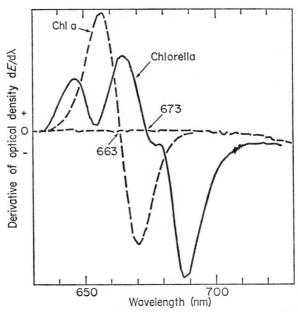

Figure 7.4. The derivative of the absorption spectrum of chlorophyll *a*, in ether solution (dashed curve), and in *Chlorella* (continuous curve). The derivatives cross the zero-line at the positions of the absorption maxima. The shoulder at about 680 nm indicates a second form of chlorophyll *a* in the alga. The dip at 654 nm represents the maximum absorption of chlorophyll *b*. From French, C. S. and H. S. Huang (1957). *Carnegie Inst. Washington, Year Book,* **56,** 267, with permission

Several notations are in use to describe these chlorophyll forms, such as C_a670, Ca_{670}, $Ca670$ and so on. Since the only form in which chlorophyll *b* is observed is C*b*650, the *a* or *b* can be omitted. The classes C670 (or C673), C680 (or C683), C695 and C705 may be compared to the forms of bacteriochlorophyll observed in the purple bacteria: B800, B850 and B890 (see Figure 7.5). In these bacteria the analogue of P700 is P890 (to be distinguished from B890). Bacteriochlorophyll absorbs at 770 nm in solution, chlorophyll at 663–670 nm in solutions, so in both cases the *in vivo* forms are displaced to the red. This may be brought about by aggregation of pigment molecules (it is known that such aggregates do have redder absorption maxima), or by increasing the hydrophobic nature of the environment (solutions in dry solvents absorb further to the red than when traces of water are present), or thirdly by complexing the pigment to protein or other non-pigment material. For the purposes of this section, in which we are concerned with structure of the thylakoid, we should note that the extent of these red-shifts, and that proportions of the various types, vary both in different species, and in different

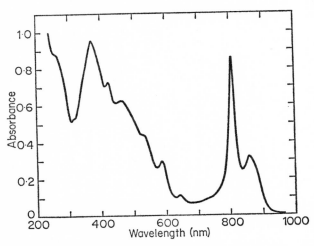

Figure 7.5. Absorption spectrum of *Rhodopseudomonas palustris* Van Niel strain 2137. Note the many peaks and shoulders, especially those at 804 nm, 858 nm and 880 nm. From Olson, J. M. and E. K. Stanton (1966). In L. P. Vernon and G. R. Seely (Eds.) *The Chlorophylls*, Academic Press, New York, p. 381, fig. 4., with permission. Copyright held by Academic Press

preparations. There may be reason to believe that the proportions of the 695–705 nm types, for example, may reflect more the kind of packing imposed on the chlorophyll by the conditions of the moment, and any alteration of the mechanical stress caused by disintegration, osmotic swelling or the forces due to the electric field believed to prevail during illumination, could alter the C695–705 material. This is the concept of 'plastic chlorophyll', which may also explain the absence of enhancement phenomena in some chloroplast preparations, as discussed in Chapter 6.

As mentioned above, at room temperature fluorescence of chloroplast material is mainly seen at 685 nm. Using the same notation, this is termed F685. At low temperatures (77°K) using liquid nitrogen, three components are usually visible, F685, F695 and F720–735. The latter is the most intense.

F685 and F695 are evoked more by illumination of system II, while F720 is stimulated by both systems. Remembering that the dichroic chlorophyll was C695–705 (absorbing system I light) and that it fluoresced at 720 nm, the lack of specificity of F720 for systems I or II at 77°K may not be significant. If the molecules come close together, the likelihood of energy migration increases inversely as the sixth power of the distance.

The specificity of F685 for system II light is much more definite; at room temperature both F685 and photoreaction II are stimulated by light absorbed by an accessory pigment, chlorophyll *b*, or, much more strikingly, phycobilins in the red and blue–green algae.

We have to identify the fluorescing forms with the C670, C680, etc. The principal rule is that the value for the *Stokes' shift* (the difference between the absorption and fluorescence maxima) is of the order of 15 nm in most chlorophyll solutions. Applying this we have:

Absorption	C(*b*)650	C670	C680	C695–705
Fluorescence	nil	F685	F695	F710–730

Action spectra (measured in terms of oxygen evolution) show that system II is deficient in pigment absorbing at more than 700 nm, and has relatively less of C680 compared with C670. In addition chlorophyll *b* is almost entirely associated with system II, although this is not always clear-cut. We know that much of the C695–705 material is part of system I, which makes it a convenient energy trap from which P700 can draw its excitation. However if the low-temperature fluorescence observation really means that some F720 (hence C695–705) is attached to system II, first there must be very little, because of the red drop, and secondly it would be a better energy trap to feed P690 if it were say C695 rather than C705. It would not matter greatly if the energy trap was at a slightly lower energy than the sink (the photochemically active pigment), provided the difference could be made up by thermal energy. (It was suggested by Goedheer that C695 might act as a de-excitation pathway for system II in dangerously strong light, as an overflow.) Figure 7.6, based on a scheme of Govindjee, Papageorgiou and Rabinowitch, represents possible associations of chlorophyll forms with the two photoreactions. The diagram does not indicate the extent to which components, say C680 of system I, are in contact between different units, either of the same system or of system II. As drawn there is none, but there may well be some contact between systems I and II (the 'spillover hypothesis').

It might be added here that experiments by Goedheer on the ability of the carotenoid pigments to sensitize fluorescence of chlorophyll *a* indicate that in the red and blue–green algae *β*-carotene transfers its energy to system

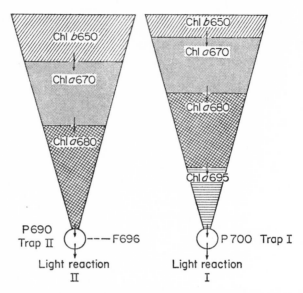

Figure 7.6. Diagram to illustrate the different distribution of the various forms of chlorophyll between the two photosystems. From Govindjee, G. Papageorgeou and E. Rabinowitch (1967). In G. G. Guilbault (Ed.), *Fluorescence*, Marcel Dekker Inc., New York, p. 511, with permission

I; in green algae and higher plants it transfers energy to both systems I and II. The xanthophylls (oxygenated carotenes) do not appear to be active at all (but note the activity of fucoxanthin in Figure 6.3).

In conclusion, we have evidence of some organization of the pigments themselves, and of at least a degree of separation of chlorophyll and carotenoid pigments into specific entities associated with the two photoreactions.

7.3 The thylakoid membrane: sub-lamellar units

The thylakoid membrane, enclosing a space, is a necessary structure for photophosphorylation according to the chemiosmotic model (see Chapter 9). Witt (1968) has argued that the whole thylakoid membrane is the 'unit' for the ion movements, on which phosphorylation is held to depend, since the uncoupler gramicidin has a half-maximal effect at a molar proportion of 1×10^5 chlorophylls. Witt calculates that 10^5 chlorophyll molecules cover an area of $2 \cdot 5 \times 10^5$ nm^2, which is the order of size of a thylakoid. However, the quantities of redox carriers such as cytochrome f are such that each electron transport chain contains 600 chlorophylls or less, about 300 in each system. Hence breakage of the thylakoid down to a size at which it could no longer seal to form the necessary vesicle for phosphorylation might still allow it to show electron transport activities.

Moderate breakage of the thylakoid does not impair any of its processes so long as the fragments can seal into ion-tight vesicles. Experiments with disintegration by sonic vibration or passage through a needle valve (the French press) followed by fractionation of the particles according to size in the ultracentrifuge have been used to identify the smallest particle that retains activity. Although various results have been reported, the more recent work (see Izawa and Good, 1965) indicates that in particles of some 2500 chlorophyll molecules (the size of the original CO_2-fixing 'photosynthetic unit' of Emerson and Arnold) the Hill reaction is seriously reduced in efficiency. The hope that by means such as this, individual active photosynthetic units could be isolated, has not yet been realized. However, it has been found possible to break the thylakoid into fragments and obtain at least a partial separation of system I and system II activities.

Michel and Michel-Wolwertz (1969) have achieved a degree of resolution by mechanical means, using a French press (see above) through which the chloroplasts were forced at a pressure of over 5 tons per square inch. They obtained three bands on a sucrose density-gradient (see Figure 7.7); the lightest band had a higher chlorophyll a/b ratio, and reduced NADP with an artificial electron donor (but less well than the original homogenate), while the other two bands had a reduced a/b ratio, and hardly reduced NADP at all. Bands 2 and 3 reduced DPIP in the Hill reaction almost twice as fast as the initial homogenate while band 1 showed very much less activity. Hence band 1 is enriched in system I and bands 2 and 3 in system II. The sizes of the particles in the bands was not determined, but there is a clear inference that there is a mechanical weakness between distinct system I or system II specific parts of the thylakoid.

Earlier, Boardman and Anderson (1965) achieved similar results, using the surface active steroid digitonin. Differential centrifugation gave a series of fractions, the lightest of which were enhanced in the ascorbate–NADP (system I) reaction (plastocyanin was required). The heavier fractions, which appeared physically much larger than the smaller particles, were more active in the Hill reaction and had a reduced ratio of chlorophyll a/b.

Application of more powerful detergents such as sodium dodecyl sulphate (SDS) (Sironval and coworkers, 1966) and sodium dodecylbenzene sulphonate (SDBS) (Thornber and workers, 1967a) produced particles small enough to be resolved by electrophoresis on polyacrylamide gels. One particle was obtained virtually devoid of chlorophyll b, and another with chlorophyll a and b approximately equal. The $(a + b)$ band migrated the faster, except for a band of free pigment. Although these small particles were inactive photo-chemically, there was strong evidence that the *a* particle was associated with the system I preparations of Boardman and Anderson, and the (*a* + *b*) particle with system II. From their small size they were regarded as complexes of protein, chlorophyll and detergent, with some lipid, and other pigments.

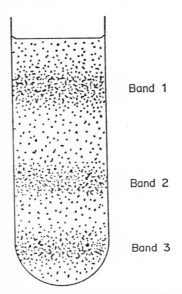

Figure 7.7. The three-banded pattern of broken spinach chloroplasts centrifuged on a sucrose gradient at 60 000 g for 45 min. From Michel, J. M. and M. R. Michel-Wolwertz (1969) *Carnegie Inst. Washington Year Book*, **67**, 508, with permission

In their amino acid composition the protein components of the particles were different. The structural protein prepared by Criddle (1966) referred to earlier in this chapter had a uniform molecular weight of some 23 000, compared with values of 150 000 obtained by Thornber for chlorophyll *a*–protein–SDS complexes. However there could be a 25 000 m.w. subunit to explain the discrepancy. The general properties of the pigment-free proteins from both procedures were in the main similar.

The neutral detergent, Triton X-100, inactivates the Hill reaction and phosphorylation at the same rate (Deamer and Crofts, 1967), and the same range of concentrations causes changes in the electron-microscopic appearance of the thylakoid membranes. Surprisingly, the ascorbate indophenol–NADP reaction after inactivation by low concentrations of Triton reappears at higher ones, and a particle has been prepared in 1 % Triton by Vernon and coworkers (1967). This particle contains 1 P700 molecule per 200 chlorophylls, and the ESR signal attributed to P700 (Treharne and coworkers, 1963) was detected when the preparation was illuminated.

The preparation of chlorophyll–protein–detergent particles does not prove the existence of chlorophyll–protein complexes in the original thylakoid, since under these conditions chlorophyll could attach itself to a variety of proteins. However the remarkable activity in System I shown by Vernon's Triton X-100 preparation is a clear demonstration that such a protein–chlorophyll complex would provide a satisfactory basis for understanding System I. Furthermore, the author observed (Gregory, Raps and Bertsch, 1971) that the SDBS–protein–chlorophyll complex I, ascribed previously to system I, was absent from a mutant of *Scenedesmus* (mutant 8 of Bishop) which was previously known to lack the reaction centre of System I. There is hence a strong case for assuming that System I depends on a specific chlorophyll–protein complex, and, by analogy, that a specific complex underlies System II also.

The experiments above in which mechanical breakage of the thylakoid resulted in some degree of separation of the photoreactions suggests that the structural segregation in the membrane of the components of either photosystem may be nearly complete. One would expect that the System I and System II regions are particles with mechanically weak connections between them. With this in mind we turn in the next section to a consideration of the evidence offered by electron microscopists for the existence of a subunit structure for the thylakoid.

7.4 Thylakoid structure by electron microscopy

The arrangement of the thylakoid lamellae within the chloroplast envelope was illustrated clearly by the electron micrograph of Plate 1 (facing p. 84). Many workers have attempted to examine thylakoid material at higher resolution in order to discern the details of any underlying subunit structure, the existence of which is made very likely by the experiments described in the previous section. In addition to the sectioning method, fragments of single thylakoid membranes can be placed flat on a specimen grid and 'negative stained' by evaporating on to them a dilute solution of a salt of a heavy metal, which settles mainly in the crevices and hollows of the membranes, so that the surface pattern can be seen in relief in the electron microscope. Thirdly, use has been made of the recent 'freeze-etching' technique, in which a living leaf is rapidly frozen, and a piece of the resulting ice-block fractured in a vacuum; water sublimes from the fracture surface, leaving the non-volatile structures of the tissue standing out in relief. A replica of this surface can be made by depositing a thin layer of metal, and this is examined in the microscope, usually in conjunction with the 'shadowing' method in which heavy metal is deposited at an angle so that the height of the relief can be judged from the length of the shadows seen on the photographs. In this way structures are revealed without the use of fixatives or stains, and there is more confidence in the fidelity of the representation.

A considerable advance was made by Park and Pon (1961) using negatively-stained fragments of thylakoids obtained by sonic disintegration and differential centrifugation. There was a clear indication of a pattern, leading to the proposal that the membrane was made of a double layer of oval particles, 10 nm by 20 nm in dimensions, which were termed 'quantasomes'. The smallest particles of the sample (which were precipitated only by centrifugation for two hours at 144 000 g) appeared to contain some six quantasomes. The same sample was analysed, and the composition of the quantasome appeared in Table 7.1 (Lichtenthaler and Park, 1963). The assumption was made that the quantasomes accounted for all the material of the thylakoid, and that there was little or no cement or matrix between them. Preparations of these particles, which contain aggregates of various sizes, have been termed 'quantasome preparations' by several workers, which may have led to some confusion; suspensions of single quantasomes were never prepared.

Further confusion arose when it was found that there were several other kinds of partical attached to, or indented into, the membrane. For example, Sironval and others (1966) observed rows of 9 nm stalked spheres attached to the edges of membrane fragments and seen in profile with negative staining (Plate 9, facing p. 84). These particles, termed oxygenomes, were seen more clearly after digitonin treatment, and were found to cover, more or less, the surface of the thylakoid. Since then the identification with System II of photosynthesis has been discarded in favour of the view of Racker that they contain enzymes concerned with the formation of ATP (see Chapter 9). (The 9 nm particles can be isolated by washing the membrane with ethylene-diaminetetraacetic acid, a chelating agent for metals such as Ca^{2+}.) In addition, the enzyme carboxydismutase also appears as particles about 10 nm in diameter, often in regions where it has an almost crystalline regularity of arrangement. The dimensions of the lattice in these regions is close to the figure given for the original quantasome pattern seen by Park and Pon, leading to the question whether the particles were attached one to each quantasome, or whether the quantasome was a confused appearance of such a regular region of carboxydismutase. This question was sidestepped when Branton and Park revised their view of thylakoid structure on the basis of their freeze-etching observations (see below).

Meanwhile, Kreutz (1966) had studied thylakoid preparations by means of low-angle X-ray scattering, and had proposed a model containing particles arranged in squares of side 4·14 nm, grouped in larger rectangles 20 nm by 14 nm (see Figure 7.8). This agreed reasonably with the current estimate of the quantasome size (18 nm by 15 nm, Park and Biggins, 1964) but the question as to which particle was actually being observed still arose.

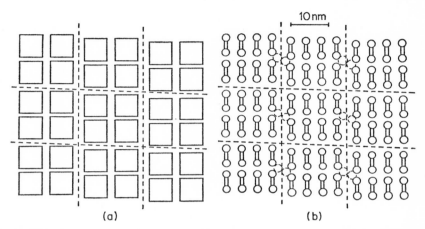

Figure 7.8. Diagrams by Kreutz combining his X-ray scattering results with the quantasome concept. The shapes are 'mass-centres' as seen from (a) the lipsid side (outside) of the thylakoid, and (b) from the protein side (inside). In T. W. Goodwin (Ed.) *Biochemistry of Chloroplasts*, Vol. 1, Academic Press, London, p. 83 with permission

Weir and Benson (1966), on the basis of their micrographs of stained sections, proposed a different subunit model (Figure 7.9) which certainly appeared to relate to the substance of the membrane itself and not to any superficial material. However it is not easy to relate their model to those obtained by the other techniques, and there is also the problem of how far fixation and staining techniques can be trusted at this extreme degree of resolution. Unfortunately none of the other methods give a cross-sectional view.

Freeze-etching studies have been actively pursued by Muhlethaler and by Branton and Park. The array of particles seen did not match those seen by other techniques, the commonest ones being 9–10 nm in diameter, and others 17·5 nm by 9 nm in size. A debate between Muhlethaler and Branton and Park occurred, concerning the plane of weakness in a frozen chloroplast along which the fracture occurred; until this was decided it was not possible

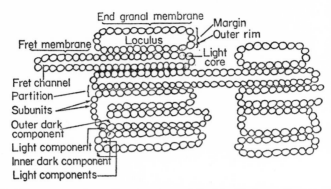

Figure 7.9. Diagram to show the structure of thylakoids in section, as deduced by T. W. Weier from electron micrographs from Weier, T. E. and A. A. Benson (1966). In T. W. Goodwin (Ed.) *Biochemistry of Chloroplasts* Vol. 1, Academic Press, London, P. 95, with permission.

to say which faces in the micrographs corresponded to which faces of the thylakoid. It is now more or less agreed that the more likely explanation is that of Branton: that the fracture plane was inside the thickness of the membrane, so that the split revealed two new faces as when one separates the two sides of a piece of toast. This view is illustrated in Figure 7.10; a striking micrograph obtained by Branton and Park is shown in Plate 11 (facing p. 84) and is interpreted (Figure 7.11) on the basis that the large (17·5 × 9 nm) particles were embedded in the inner part of the membrane, and the smaller

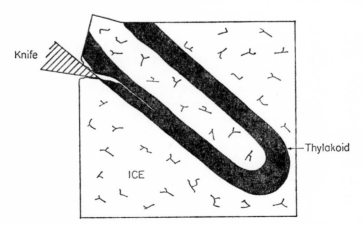

Figure 7.10. Diagram illustrating the principle of freeze etching, and of the splitting of thylakoid membranes inside their thickness

ones in the outer section. They claim that the larger particles are only found in the granum regions, and not in the inter-grana thylakoids. They further claim that the larger particles are concerned with system II; there are reports (Woo and coworkers, 1970) that plants which contain granulate and agranulate chloroplasts do not have System II (oxygen evolution) in the latter. Hence the term 'quantasome' is dubious, and may be soon discarded if the new two-particle concept becomes established. It is clearly more in accord

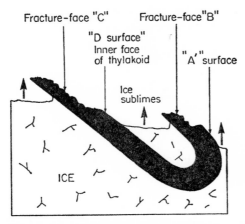

Figure 7.11. An interpretation of Plate 11 shown as resulting from the situation of Fig. 7.10.

with the two-light-reaction concept, and with the results of the previous section, that two distinct fundamental sub-lamellar particles should exist, and it seems also reasonable that a third entity, a matrix, should exist in keeping with the structures of other cell membranes.

7.5 Summary

Thylakoid membranes of chloroplasts contain proteins, lipid and pigments. Much of the protein is 'structural protein', and there are probably two principal kinds. Some of the chlorophyll is attached to the protein and some appears to be free; the different physical situations for chlorophyll in the thylakoid appear to be the cause of the variations in the absorption maxima, of which the principal types are C(a)670, C680, C695 and C705. Only the long wavelength forms are oriented. The various forms of chlorophyll a, and chlorophyll b, contribute energy in different proportions to the two light reactions of photosynthesis.

Thylakoid membranes can be broken down to give fragments which may result in a greater or lesser degree of functional separation of the two photo-reactions, or of components associated with particular photoreactions if activity is lost. System II is more easily inactivated than System I. The thylakoid membrane contains two kinds of particle embedded within it which may be the basis of the above separations, and provide a structural rationale of the two-light-reaction theory. An earlier and widely-used term, the 'quantasome', referred to a hypothetical single type of subunit, the significance of which must be regarded as doubtful. The surface of the thylakoid carries other particles such as ATPase 'coupling factors' and the enzyme carboxydismutase which must be distinguished from the internal particles in the interpretation of micrographs.

References

Boardman, N. K. and J. M. Anderson (1964). *Nature*, **203**, 166.

Criddle, R. S. (1966). In T. W. Goodwin (Ed.), *Biochemistry of Chloroplasts*, Vol. 1, Academic Press, London, p. 203.

Deamer, D. W. and A. R. Crofts (1967). *J. Cell Biol.*, **33**, 395.

French, C. S. and H. S. Huang (1957). *Carnegie Inst. Washington Year Book*, **56**, 267.

French, C. S. and L. Prager (1969). In H. Metzner (Ed.), *Progress in Photosynthesis Research*, Institut fur Chemische Pflanzenphysiologie, Tubingen, p. 555.

Govindjee, G. Papageorgeou and E. Rabinowitch (1967). In G. G. Guilbault (Ed.), *Fluorescence*, Dekker, New York, p. 511.

Gregory, R. P. F., S. Raps and W. F. Bertsch (1971). *Biochem. Biophys. Acta* (In press).

Izawa, S. and N. E. Good (1965). *Biochim. Biophys. Acta*, **102**, 20.

Kreutz, W. (1966). In T. W. Goodwin (Ed.), *Biochemistry of Chloroplasts*, Vol. 2, Academic Press, London, p. 83.

Lichtenthaler, H. K. and R. B. Park (1963). *Nature*, **198**, 1070.

Michel, J. M. and M. R. Michel-Wolwertz (1969). In H. Metzner (Ed.), *Progress in Photosynthesis Research*, Vol. 1, Institut für Chemische Pflanzenphysiologie, Tubingen, p. 115.

Muhlethaler, K. (1966). In T. W. Goodwin (Ed.), *Biochemistry of Chloroplasts*, Vol. 1, Academic Press, London, p. 49.

Olson, J. M. and E. K. Stanton (1966). In L. P. Vernon and G. R. Seely (Eds.), *The Chlorophylls*, Academic Press, New York, p. 381.

Olson, R. A. (1963), In *Photosynthetic Mechanisms of Green Plants*, Publication 1145, NAS-NRC, Washington, D.C., p. 545.

Park, R. B. and J. Biggins (1964). *Science*, **144**, 1009.

Park, R. B. and D. Branton (1967). In *Energy Conversion by the Photosynthetic Apparatus*, Brookhaven symposium No. 19, Brookhaven National Laboratory, New York, p. 341.

Park, R. B. and N. G. Pon (1961). *J. Molec. Biol.*, **3**, 1.

Sironval, C. and coworkers (1966). In J. B. Thomas and J. C. Goedheer (Eds.), *Currents in Photosynthesis*, Donker, Rotterdam, p. 111.

Thornber, J. P., R. P. F. Gregory, C. A. Smith and J. L. Bailey (1967). *Biochemistry*, **6**, 391.

Thornber, J. P., J. C. Stewart, M. C. W. Hatton and J. L. Bailey (1967). *Biochemistry*, **6**, 2006.

Treharne, R. W., T. E. Brown and L. P. Vernon (1963). *Biochim. Biophys. Acta*, **75**, 324.

Vernon, L. P., E. R. Shaw and B. Ke (1966). *J. Biol. Chem.*, **241**, 4101.

Weier, T. E., and A. A. Benson (1966). In T. W. Goodwin (Ed.), *Biochemistry of Chloroplasts*, Vol. 1, Academic Press, London, p. 91.

Witt, H. T. (1968). In H. Staudinger and B. Hess (Eds.), *Biochemie des Sauerstoffs*, Springer, Berlin.

Woo, K. C. and coworkers (1970). *Proc. Natl. Acad. Sci. U.S.* **67**, 18.

CHAPTER 8

Photosynthetic electron transport

In this chapter we shall examine the evidence that certain redox materials in thylakoids and bacterial chromatophores are oxidized and reduced during photosynthesis, and the means by which such evidence is obtained. From there we can then discuss sequences of redox carriers (electron transport chains) at various key places. These partial sequences can be incorporated into more than one type of model, and some of the models currently proposed will be compared to the familiar zig-zag scheme.

8.1 Evidence for the participation of certain carriers in photosynthetic electron transport

8.11 Quinones

Plastoquinone (PQ) is the most abundant redox material in the chloroplast, its molar ratio to chlorophyll being 1:7 of which some 80% is PQ-A, 20% PQ-C. This and its physical properties (see section 4.51) enabled quantitative extraction to be performed after various treatments so that changes in the redox state could be established. After illumination it was shown that a part of the PQ pool was reduced, although the reported degree of reduction varied rather widely. However it must be borne in mind that approximately half of the PQ of the chloroplast occurs in osmiophilic globules or plasto-globuli which may be metabolically inert and that these globules increase in size during the life of the leaf. It appears reasonable that most of the PQ of the thylakoid membrane can be reversibly oxidized and reduced during light–dark cycles.

Further evidence was obtained by Henninger and Crane (1966) that the Hill reaction with indophenol as oxidant was inhibited by exhaustive extraction of PQ with hydrocarbon solvents. Restoration of the activity required both PQ-A and also PQ-C in approximately their *in vivo* proportions. PQ-B may possibly act at the same site as PQ-C, but it is difficult to isolate and test.

126

Thirdly, spectroscopic investigations have revealed on illumination changes at 255 nm, which are indicative of quinones of this series (Klingenberg and coworkers, 1962); this method also allows determination of the kinetics, which is the necessary supplement to the foregoing evidence if it is to be decided whether the quinones are on a direct line of electron transport or a side-line. Thus Witt, Suerra and Vater (1966) illuminated chloroplasts with 'system I' light (700–730 nm) followed by a flash of shorter wavelength. They observed a 255 nm change with a time constant* of 10^{-5} s, indicating the reduction of PQ by system II, followed by reversal in 10^{-2} s. While this last change is slow, it is consistent with the early kinetic experiments of Emerson and Arnold (1932) which showed that flashes of (neon) light were as effective as continuous light if the repetition rate was 10^{-2} s or faster.

It seems to be established, therefore that plastoquinone is directly concerned with electron transport, that it is reduced by 'system II' light, and that PQ-A and PQ-C are both required.

Ubiquinone (UQ), which is the principal quinone taking part in electron transport in mitochondria, is found in the photosynthetic bacteria. Its role in photosynthesis is established by absorption changes at 270–275 nm, and by the demonstration that photophosphorylation is blocked in chromatophores from which UQ has been extracted, and is restored by the specific readdition of UQ. Plastoquinone is not found in the bacteria.

There is some variation in the isoprenoid side-chains of these quinones, but the commonest forms are PQ-9 in green plants and UQ-10 in bacteria.

Other quinones such as tocopheryl quinones occur with their reduced (tocopherol) forms, so that a similar function remains possible. Vitamin K_1 (a naphthoquinone derivative) has been shown to be less abundant in algae when they are grown non-photosynthetically, but there is no indication of its function.

8.12 Proteins

Plastocyanin. This protein is extracted from the thylakoid membrane by ultrasonic disintegration or detergent treatment, and sublamellar particles prepared by these methods are unable to carry out the reduction of NADP (via ferredoxin) with reduced indophenol as the reductant. Addition of plastocyanin restores the activity. The 598 nm absorption peak of the cupric copper in plastocyanin, while relatively intense, is so wide that spectroscopic observations of reversible redox changes are difficult to obtain. The inhibitor salicylaldoxime was thought to react specifically with plastocyanin

---

* The time constant of a reaction is an indication of its duration: for an exponential decay, the time constant is the time taken for 63 % $(1 - 1/e)$ of the total change to occur.

copper, but the overall effects are not consistent with the readdition studies mentioned above.

Rubimedin is a red protein extracted by water from dry, heptane-extracted chloroplasts, which is reported to be a redox carrier on the basis of readdition studies. Henninger and Crane (1967) found that it was an alternative to the plastoquinones, at least in part, in that it restored activity to heptane-extracted chloroplasts. There is little other information available at the present time.

Cytochromes. Spectral changes associated with reduction of cytochrome f are well documented both in the α-region at 555 nm and the γ-region at 407 nm. The chloroplast is relatively opaque, due to the high concentration of pigment, even in the region 540–570 nm. However the sharp spectral peaks of cytochromes allow kinetic analysis to be made *in situ*, with modern sensitive spectrophotometers. The results of, for example, Witt, Suerra and Vater (1966) indicate that cytochrome f is reduced by system II light with a time constant of 10^{-2} s and is oxidized in 10^{-4} s. These changes are affected by uncouplers of phosphorylation. Cytochrome f is present in chloroplasts, at a concentrational 1:430 total chlorophyll on a molar basis (Davenport and Hill, 1952; Anderson, Fork and Amesz, 1966). Cytochrome b_6 has been estimated by several groups, and is approximately twice as abundant as cytochrome f. By spectroscopy at 412, 435 and 563 nm, the reduction of cytochrome b_6 has been observed, both with system II light (Rumberg, 1966) and with system I light (Hind and Olson, 1966). Kinetically, Rumberg found that the time of reduction by system II light was 10^{-2} s, but the oxidation was slower (10^{-1} s) than required by the zig-zag hypothesis. Both cytochromes f and b_6 are present in the approximate proportions of 1–3 per photosynthetic unit but, while a good case can be made for the involvement of cytochrome f, the role of cytochrome b_6 is obscure.

Observations of changes at 559 nm are more difficult to identify. Cytochrome b_3, prepared from pea microsomes, has a potential $E'_{0,7}$ of between 58 and 260 mV; Lundegardh (1962) identified this cytochrome in chloroplasts on the basis of its absorption maximum at 559 nm, but did not measure the potential. Bendall (1968) measured a potential of $+0.37$ V for a 559 nm band by titration with ferricyanide, and concluded that it revealed a b-type cytochrome. Hind and Olson (1966) observed a 559 nm band in a sub-lamellar particle, prepared using the detergent Triton X-100, which had system I activity. This band was only visible at 77°K, and the authors commented that it might be a hitherto unreported splitting of the b_6 α-band. Bendall (1968) has reported such a splitting of the b_6 α-band at 563 nm into 559 and 563 nm components at 77°K. Both appear to have E'_0(pH 7) $= 0.0$ V.

There are mutant non-photosynthetic strains of the alga *Chlamydomonas* (Ac 141, Ac 115 and F34 of Levine) lacking the b_{559} component (potential not recorded) (Levine 1969).

The cytochrome b-559 occurs in approximately equal proportions to cytochrome b_6. It should be borne in mind that b cytochromes are properly defined in terms of a protohaem-IX prosthetic group that can be detached by acid acetone. At present the 559 nm material, at least that observed by Bendall (1968) is better referred to as cytochrome 559.

In bacteria, the cytochromes found are different in each species. We shall be mainly concerned in this chapter with work done on the purple bacteria

Table 8.1. Cytochromes in two purple bacteria

Cytochrome	Synonyms or absorption maxima	$E'_{0.7}$	
Chromatium			
C552	(C553, C423·5)	10 mV	possesses FMN
C555	(C556, C422)	340 mV	
cc'	(RHP, C430)	−5 mV	
R. rubrum			
c_2	550, 415	338 mV	
cc' (RHP)	550, 423	−8 mV	
b	560, 430	0	

Chromatium (Thiorhodaceae, anaerobic) and *Rhodospirillum rubrum* (Athiorhodaceae, a facultative aerobe). It will be seen (Table 8.1) that b cytochromes are not found in *Chromatium*. Furthermore although cytochrome b is often placed on photosynthetic pathways on the grounds of 430 nm spectral changes, Bartsch has suggested that a better identification might be cytochrome cc'. Cytochrome b itself may be associated more with aerobic respiration than with photosynthesis.

Ferredoxin-reducing substance. Trebst has reported experiments in which antibodies to thylakoid material were prepared in rabbits. The antigenic substances were found to be the ATPase 'coupling factor', the enzyme ferredoxin–NADP oxidoreductase, and an unknown component. The antibody blocked the reduction of ferredoxin by photoreaction I in chloroplasts. It was active against a soluble preparation obtained by San Pietro in a different laboratory, and this preparation would relieve the inhibition caused by the antibody. The preparation of San Pietro (ferredoxin-reducing substance, FRS) possessed heat labile and heat-stable components. There has long been discussion of pteridines as redox intermediates in the position of FRS; this possibility is currently being examined.

Ferredoxin–NADP oxidoreductase, EC.1.6.99.4. In non-cyclic electron transport in green plants there can be little doubt that ferredoxin is required for the reduction of NADP, and that the reduction of NADP by reduced ferredoxin is catalysed by the flavoprotein enzyme ferredoxin–NADP oxidoreductase, although it is possible that other flavoproteins may be present able to catalyse this reaction to some extent. The reductase has been isolated from spinach (Shin and coworkers, 1963), and its role in the thylakoid has been shown by San Pietro (Keister, San Pietro and Stolzenbach, 1960) who applied an antibody specific to the reductase which inhibited the reduction of NADP. Forti and Zanetti (1969) has claimed that this flavoprotein has another site of action in addition to that of transferring electrons from ferredoxin to NADP: a complex can be formed between the flavoprotein and cytochrome *f*, and that *in the absence of oxidized NADP* washed chloroplast lamellae will reduce exogenous cytochrome *f* in such a complex. They argue that this complex exists *in vivo*, and provides a pathway for cyclic electron flow:

System I—ferredoxin—reductase—cyt f—System I

On this theory cyclic flow would only take place when all the NADP was reduced, as would happen if the carbon pathway became deficient in ATP.

Ferredoxin. This is well established both as the 'natural' Hill oxidant and as the natural factor for cyclic electron flow (as opposed to artificial reagents such as PMS). One criticism, that reduced ferredoxin is so readily oxidized by oxygen as to present a serious leak, appears to be met by the great affinity for ferredoxin of the ferredoxin–NADP oxidoreductase. A second criticism, that the concentration required for cyclic photophosphorylation in broken (stroma-less) chloroplasts exceeds the normal chloroplast concentration, awaits precise measurement of the latter value. It has been very clearly demonstrated that ferredoxin is reduced by chloroplasts (using the ascorbate–indophenol donor system, and the system II inhibitor DCMU to prevent the production of oxygen from reoxidizing the ferredoxin) and that reduced ferredoxin will, given the NADP reductase, bring about the reduction of NADP. In bacteria there is a different NAD-reduction pathway, since although ferredoxin has been detected in all three groups of photosynthetic bacteria, it has not been shown to be necessary for cyclic photophosphorylation, nor is it always necessary for NAD reduction. However the reductive carboxylations of acetyl coA and succinyl CoA (see section 5.22) which shed new light on the bacterial carbon metabolism, do require reduced ferredoxin.

It will be argued below that at least in *Chromatium* there are two light-driven non-cyclic electron transport chains, one of which will reduce ferredoxin and the other NAD directly.

8.2 Sequences of redox carriers in electron transport in green plants

Given that there are two light reactions driving redox reactions, there must be a primary oxidant (X) and reductant (Y) for each. These symbols represent unknown intermediates, that can be investigated kinetically. It is probably true to say that most of the work reported in this section has been done and presented on the basis of the series (zig-zag) formulation of the two-quanta hypothesis. A separate section will be given to assessing the merits of the zig-zag theory against for example that of Arnon (1966).

8.21 System I

The site X_I. Reference has been made (section 6.21) to the discovery of artificial redox couples with very negative standard potentials that could be reduced by chloroplast lamellae with system I light. By allowing the process to reach equilibrium, the actual potential of the electron source could be calculated from the observed redox level of the material added. Thus Zweig and Avron found a potential of approximately -0.5 V at pH 7.8, Kok reported -0.65 V, and Black -0.52 V. The question arises whether the electron was expelled from chlorophyll directly, or via an intermediate. There was no direct demonstration of any intermediate, but on kinetic grounds, if the electron is to be accepted by a freely diffusible molecule (ferredoxin or exogenous oxidants) the reaction must be relatively slow, requiring the excited electron to have a long life, which is difficult to believe. A bound intermediate, such as FRS, would store the electron for a longer period. There is a flavoprotein, phytoflavin, that in iron-deficient plants may replace ferredoxin to some extent (see Bothe, 1969). Pteridines have been the subject of some speculation and may be reduced by system I (see Fujita and Myers, 1967). Nevertheless the topic remains speculative unless (to quote Arnon) 'a stronger electron carrier is actually isolated from the photosynthetic apparatus'.

We require a mechanism for the transfer of the electron from chlorophyll to the first acceptor, perhaps FRS. This may be visualized in two ways if one accepts that P700 is a genuine chlorophyll photoproduct equivalent to $\cdot Chl^+$ observed in photochemistry of chlorophyll *in vitro*. The first mechanism (1) is a Rabinowitch reaction (see Chapter 3) except that we have to postulate that the acceptor is bound to the thylakoid to account for the high speed of the reaction. (Witt has measured the half-time of appearance of oxidized P700 as less than 2×10^{-8} s.)

$$\begin{array}{cc} \text{Chl}^* & \text{Chl} \\ \text{P700—FRS} \rightarrow \cdot\text{P700}^+\text{—FRS}^- \end{array} \qquad (1)$$

In the second mechanism (2) we have the formation of both the radical–cation and the radical–anion of chlorophyll; the latter has been shown to be an intermediate in the Krasnowskii reaction (see Chapter 3).

$$\begin{array}{cc} \text{Chl}^* & \text{Chl} \\ \text{P700—Chl} \rightarrow \cdot\text{P700}^+\text{—}\cdot\text{Chl}^- \end{array} \qquad (2)$$

$$\cdot\text{Chl}^- + \text{FRS} \rightarrow \text{Chl} + \text{FRS}^-$$

Either mechanism would account for the ESR (electron spin resonance) signal observed in connection with photoreaction I.

To summarize, a reasonable scheme for electron transport from the site X_I to NADP is as follows:

$$\text{(System I)} \rightarrow \text{FRS} \rightarrow \text{ferredoxin} \rightarrow \text{reductase} \rightarrow \text{NADP}$$
$$? \searrow \text{(phytoflavin)} \nearrow \qquad (3)$$

The site Y_I. The absorption band visible in chloroplasts at 700 nm, and that is bleached either by illumination, or by oxidizing agents such as ferricyanide, is termed P700, and is believed to be a specific chlorophyll a molecule modified by its environment, although it has not been isolated. The bleaching is presumed to be oxidation, and the calculated standard potential is 0·43 V. The oxidation is driven by system I light, and hence P700 is a candidate for the role of Y_I. Kinetically, the oxidation is fast, and no component has been shown to be oxidized faster.

Reduction of P700 can be brought about by illumination with system II light, or by injection of electrons from exogenous N,N,N′,N′-tetramethyl-enediamine (TMPD), or from the ascorbate–DCPIP couple, in which case there is a requirement for plastocyanin. Cytochrome f is oxidized synchronous-ly with the reduction of P700 (Witt, 1968). Exogenous plastocyanin has been shown to accelerate this process, presumably because it tends to leach out of thylakoids in water. A mutant of *Chlamydomonas* (Ac 208: Gorman and Levine, 1966), lacking only plastocyanin was only able to oxidize cytochrome f when plastocyanin was added to chloroplast fragments.

Avron and Chance (1966) have observed that there is a probable phosphoryl-ation site between system II and cytochrome f. (See section 9.2.) There is some doubt whether the reduction of NADP by the ascorbate–DCPIP couple involves phosphorylation; the observation by Avron and Chance that cytochrome f did not appear to be on the direct pathway for this reaction would suggest a scheme such as (4):

$$e^- \longrightarrow Cyt\,f \longrightarrow Pcy \longrightarrow P700 \longrightarrow System\ I \qquad (4)$$

$$\qquad\qquad \uparrow \qquad\qquad \uparrow$$

$$(ascorbate \longrightarrow DCPIP) \quad (TMPD)$$

Scheme (4) separates cytochrome f from P700. This is apparently contrary to the observations of Amesz and Fork, (1967) who showed kinetically that four molecules of cytochrome f were intimately associated with one molecule of P700 in the active centre of system I. (This was based on the assumption that the absorption of P700 is of the same magnitude as chlorophyll a.) They noted however that plastocyanin has a potential very close to that of cytochrome f, and possibly plastocyanin was involved in the active centre also; this is the case represented in scheme (4).

8.22 System II

Kok and Cheniae (1968) refer to system II as the 'inner sanctum' of photosynthesis, since compared with details concerning the sequences around system I, system II remains virtually totally obscure. The point is emphasized by the finding (see Chapter 7) that the Hill reaction needs comparatively large sections of the thylakoid, possibly of a size sufficient to make sealed vesicles, whereas 'system I particles' are small, containing one 'set' of chlorophyll molecules and redox compounds.

The site Y_{II}. The intermediate Y_{II} must be of sufficiently high potential to generate oxygen from water, that is, higher than $+0.82$ V for 1 atm. oxygen, at pH 7. The potential required declines by 0.06 V per pH unit, so that if the effective pH at the oxygen-evolving site were greater than neutrality, Y_{II} could have a lower potential than $+0.82$ V. However, the reaction

$$2H_2O \rightarrow O_2 + 4H^+ + 4e^-$$

produces hydrogen ions, and another source of energy would be required to remove them and so maintain the elevated pH. An interesting hypothesis can be constructed, in which one light reaction maintains a high pH by hydrogen ion pumping, and the second oxidizes water via a low-potential Y_{II}. In the absence of experimental support,* however, this scheme can be left as an exercise in model building, and the search for a high-potential Y_{II} continued.

* See section 9.2 (p. 158) where it is shown that the inside of the thylakoid does in fact become acid (pH 5·1) during photosynthesis.

There is at present no indication of what sort of material (or materials) Y_{II} might be. Haem compounds such as the enzyme catalase (EC.1.11.1.6) which decomposes hydrogen peroxide to oxygen and water might seem a likely candidate for Y_{II}, but is discounted first on the grounds that inhibitors such as the triazines have different effects on catalase and on the Hill reaction, and secondly the production of hydrogen peroxide from water that would have to precede a catalase-catalysed step is energetically more difficult than the evolution of oxygen directly. (Cytochrome f has a remarkably vigorous catalase activity, which must be considered irrelevant.)

Manganese has been suggested as a possible Y_{II} component. Since the work of Pirson and of Kessler, it has been known that manganese-deficient algae were impaired in their photosynthetic activity, and it was later shown that this impairment lay entirely within system II. Kok and Cheniae (1967) have reviewed this topic; they consider that while there is no evidence of a valency-change in the manganese of the thylakoid during photosynthesis, nevertheless the Hill reaction does depend on the presence of strongly bound manganese. Protein is likely to be the ligand, but the valency state of the metal is not known. We may note that the high potential of the manganic manganous ion couple (Mn^{3+}/Mn^{2+}, $E_0' = 1 \cdot 5$ V, pH7) offers an attractive basis for theorizing. Growth of algae under conditions of manganese deficiency causes them to lose their photosynthetic activity, and they live heterotrophically. The deficiency lies in system II. The chloroplasts of these algae are, however, still able to carry out a form of electron transport in which hydroxylamine (NH_2OH) acts as an electron donor to system II, given an acceptor such as ferricyanide. It seems as if hydroxylamine by-passes a manganese-requiring step. This is prevented by severe manganese deficiency, so possibly there is more than one site in system II where manganese is needed. The thylakoids of these deficient algae are grossly malformed, and this may provide part of the explanation; nevertheless it is observed that addition of manganese to a deficient culture causes photosynthetic ability to be regenerated in a matter of minutes, which might be thought to preclude protein synthesis or membrane formation.

The chloride ion is also required for system II activity, and chloride deficient chloroplasts behave in the same way as those deficient in manganese, as has been shown by Izawa, Heath and Hind (1969). In this case addition of chloride has an almost instantaneous restoring action. A third inorganic requirement for system II is carbon dioxide. This is quite independent of its metabolic fixation by stroma enzymes. It is possible that it acts, like manganese and the chloride ion, at site Y_{II}. (This effect of carbon dioxide is fundamental to the theory of Warburg, see section 8.23, in which photosynthetic oxygen evolution comes directly from activated carbon dioxide with the one-step formation of carbohydrate.)

There are several indirect lines of evidence which shed some light on the electron transport pathway between chlorophyll and oxygen. Witt's group have measured the half-time of the water-splitting reaction, by measuring the dependence of its rate on the frequency of flashing light following the same principle as the Emerson–Arnold experiment. In this case chloroplasts when supplied with dichlorophenol–indophenol as an electron acceptor showed a rate-limiting step in the Hill reaction of 2×10^{-3} s. It was already known that the rate of reduction of plastoquinone by photoreaction II was faster than this, hence this limit of 1×10^{-3} s must belong to a reaction on the Y_{II} side of system II. Since then, Witt's group have observed a change at 682–690 nm which has a rise-time of less than 2×10^{-8} s, and a decay-time of 2×10^{-4} s. They regard this as a photochemically active form of chlorophyll a belonging to system II (they term it Chl-a_{II} or P690). If this identification is correct, then Y_{II} must presumably be oxidized (in a saturating flash) with a half-time of 2×10^{-4} s, and reduced by water in 2×10^{-3} s.

Some remarkable experiments by Joliot, Barbieri and Chabaud (1969) and by Kok, Forbush and McGloin (1970) indicate an electron storage system in the Y_{II} position. In one of the experiments the yield of oxygen in a bright short flash is measured, then a second flash is given and the yield again determined. This is repeated some thirty times. Provided that the material

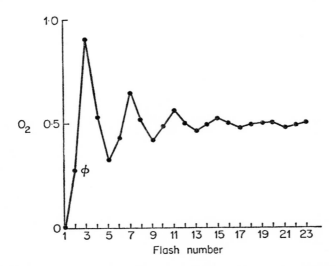

Figure 8.1. Amount of oxygen released by each of a series of flashes (of saturating intensity) from cells of *Chlorella* previously kept in the dark for 3 min. The dark interval between flashes was 300 ms. From Joliot, P., G. Barbieri and R. Chabaud (1969) *Photochem. Photobiol.* **10**, 309, Pergamon Press, with permission

had been kept in darkness for half an hour or longer, these workers were able to demonstrate that the yield of oxygen was periodic (Figure 8.1), being zero for the first flash and reaching maxima on the 3rd, 7th, 11th flashes and so on. This indicates clearly that there is a store of positive charges, filled one at a time by system II and emptied four at a time by the reaction

$$4OH^- \rightarrow O_2 + 4H^+ + 4e^-.$$

To account for the high yield on the third flash, Joliot suggested that partially filled sites could cooperate to form oxygen

$$[O] + [O] \rightarrow O_2$$

by a switch mechanism. Kok has suggested that if the store S passed through the sequence

$$S_0 \xrightarrow{h\nu} S^+ \xrightarrow{h\nu} S^{2+} \xrightarrow{h\nu} S^{3+} \xrightarrow{h\nu} S^{4+} \longrightarrow S_0 + O_2$$

the phenomena could be explained if S_0 and S^+ were equally stable, even during long periods (30 minutes) of darkness. In Kok's view (which appears to be more attractive at the present time) each store functions independently of the others, and collects one electron at a time.

System II is responsible for most of the fluorescence of chlorophyll at room temperature. It is more vulnerable to heat, fragmentation and inhibitors than system I, and, if the P690 hypothesis of Witt is correct, the photochemical centre retains its energy for a relatively long time (2×10^{-4} s). All these considerations point to the chlorophyll of system II being organized in a more complex manner compared with that of system I. Bertsch (1969) has offered a model for the operation of system II based on his measurements of the Strehler–Arnold delayed light emission. This is a weak light emission, about 1% of the amount of fluorescence, which in turn may be some 1% of the incident light, emitted from the singlet excitation state of chlorophyll (680–720 nm). This time course of the emission is of two parts: the first part is characterized by a rapid decay, the rate of which is scarcely affected by temperature, and a second part which persists much longer, and which decays faster at higher temperatures. The kinetics of the decay of these delayed light emissions are complicated so that the evaluation of the half-life has no precise meaning, but it is striking that the non-thermal (fast) part is conveniently observed in the interval $0 \cdot 5$–5×10^{-3} s, while the thermal part can be observed for maybe a hundred times longer, and indeed in a different apparatus can be shown to persist for many minutes. Both emissions come from system II; this is established by the enormously-diminished emission from *Scenedesmus* mutant 11 of Bishop, which has no functional system II. (The emission from bacteria is likewise several orders of magnitude less than that from green plants.) Bertsch envisages a region in which some chlorophyll *a* molecules form a semiconductor lattice; while intrinsically an insulator, such a system will allow the efficient migration of electrons and holes (positive charges) injected into it. This could occur when an exciton arrived, and the electron–hole pair would migrate to a trap, where one of the pair is held and the other remains close to it. (See Figure 8.2.) When electron transport operates, the trapped charge is taken up by X_{II} or Y_{II}, leaving the associated member of the original pair free to wander. There would be equal numbers of electron traps and hole traps, and hence when system II electron transport was taking place there would be similar numbers of positive and negative wandering charges in the semiconductor matrix. Some of these could neutralize each other with the emission of light. It is in fact experimentally observed that the intensity of the fast non-thermal delayed light is greater, the greater the electron transport activity of system II. Inhibitors such as DCMU which specifically inhibit photoreaction II eliminate the non-thermal delayed light, leaving the thermal component.

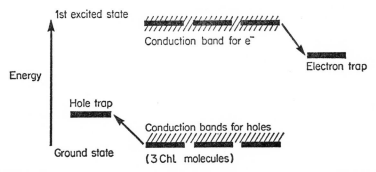

Figure 8.2. A model for the reaction centre of system II in terms of a semiconducting group of chlorophyll molecules, after W. F. Bertsch.

The thermal component of the delayed light is thought of as due to the re-excitation of a trapped charge back into the conduction band of the semiconducting chlorophyll. Work must of course be done to prise the charge out of its trap, and this is taken from the thermal movement of the molecules. Hence an increase in temperature both brightens and shortens the thermal part of the emission. Furthermore, the Arrhenius equation relates the velocity of a reaction, k, to the energy of activation, E_A:

$$k = A \cdot e^{-E_A/RT}$$

where A is a constant and R and T, have their usual significance. Solving this for the observed kinetics of the thermal delayed light gives an activation energy of 23 kcal mole^{-1}, or 1 eV. Now the energy gap between the positive and negative conduction bands of the chlorophyll must be related to the wavelength of the emission, about 700 nm, giving 1·7–1·8 eV. If the trapping energy is about 1 eV, then the energy gap between the positive and negative traps is of the order of 0·8 V. This is an approximation to the difference in the potential of Y_{II} and X_{II}, and appears plausible.

The traps in Bertsch's theory may be reaction centre chlorophyll molecules, or they may be other redox materials in electronic contact with the semiconducting pigment. The relationship of the proposed traps to the P690 concept of Witt is not defined. One should note however that the non-thermal delayed light emission decays at a very similar rate to the decay of P690, and the thermal emission, persisting for many minutes, reminds one of the long period of darkness necessary to deplete Y_{II} of its positive charges in the periodic-oxygen-pulse experiments of Joliot.

Recent results from several laboratories have shown that the delayed light emission from chloroplasts can be influenced by conditions such as the presence of salts or the 'high-energy state' (precursor of ATP). Such conditions are also known to affect fluorescence emission spectra, and it may be that osmotic forces, or the forces associated with phosphorylation, alter the degree of aggregation of the chlorophyll molecules. The resulting changes in the capacity of the energy-storing system could account for the observed effects on fluorescence and the delayed light emission.

There is a remarkable claim by Knaff and Arnon (1969) that cytochrome 559 is oxidized by system II. This is unexpected, since although cytochrome 559 accompanies system II in digitonin-fractionation experiments, it has an inappropriate potential (0·37 V) to fit the zig-zag scheme at either X_{II} (about 0 V) or to decompose water at pH 7 (0·82 V). Nevertheless these authors show clearly that a 559 nm cytochrome is oxidized by system II light at 90°K, suggesting that the material is in close electronic contact with the reaction centre. At room temperature the effect is not seen unless the Hill reaction is inhibited by extraction of the chloroplasts with 0.8 M tris buffer. (Under these conditions hydroxylamine and other reagents can still donate electrons to system II and thence to a Hill reagent, so the reaction centre of system II is not damaged.) Knaff and Arnon conclude that cytochrome 559 is an intermediate in the pathway between water and photoreaction II. This conclusion, the zig-zag scheme of photosynthetic electron transport, and the potential of 0·37 V measured for cytochrome 559 cannot all be maintained; Arnon had already rejected the zig-zag scheme, but his alternative still required evolution of oxygen from water at pH 7, that is, 0·82 V. (It should be noted that it is no help to criticize the identity of the 559 nm band. Although b_6 does split to give a 559 nm component at low temperature, the lower potential, 0·0 V, makes the problem worse, and in any event there is evidence that cytochrome b_6 is reduced and not oxidized by system II.)

Arnon's solution (Knaff and Arnon, 1969) is to place cytochrome 559 on a pathway connecting *two* system II light reactions (system IIa and system IIb). These would have more or less indistinguishable action spectra. Since system I is still required for photophosphorylation, Arnon is introducing here a three-light-reaction hypothesis.

To summarize, we may say that there is a chemical entity at the Y_{II} site, that collects positive charges from system II and generates oxygen from water in a four electron-reaction. Manganese is a part of the electron pathway from water, but with other donors this metal is less important. Chloride ions, and possibly carbon dioxide (or bicarbonate ions), have an activating effect close to this site. At pH 7 Y_{II} would need to have E_0' exceeding 0·82 V. No substance is known to fit this specification.

The site X_{II}. Witt's group have made good progress in this sequence as well. Using the rapid-reaction spectroscopic technique they have identified a spectral change operated by system II at 320 nm, rising in less than 2×10^{-8} s (the limit of the apparatus) and decaying in 6×10^{-4} s. The decay time is the same as the rise time of plastoquinone (reduction) observed at 263 nm, and they consider the 320 nm-peak to be the semiquinone form of plastoquinone. (Semiquinones are unstable free radicals formed by a one-electron reduction of a quinone or oxidation of a quinol.) The rise time of 2×10^{-8} is as fast as any change they observe, and with their present equipment it is at least as fast as the oxidation of P700 or P690. This suggests that PQ is in close electronic contact with the chlorophyll of system II. PQ may be the electron trap represented in Bertsch's model, or photoreaction II may be described by

$$\text{Chl} \xrightarrow{\text{Chl}^*} \text{PQ} \rightarrow \cdot\text{Chl}^+ \xrightarrow{\text{Chl}} \cdot\text{PQ}^-$$

which is virtually the same thing, but allows the active centre to be a single chlorophyll molecule analogous to P700 in system I. Witt and his colleagues are able to measure the quantity of PQ from the kinetics of the reduction. They suggest that there are 3–5 molecules of PQ per electron transport chain, hence the ratio of active PQ to chlorophyll is about $\frac{1}{150}$. They do not specify which PQ is operating, but both PQA and PQ-C are sufficiently abundant to account for all of this total.

Analysis of fluorescence data gives a somewhat different picture. At room temperature system II produces almost all of the fluorescence of chloroplasts, but the intensity of the emission depends on the treatment that the preparation has received. Thus after a period of darkness in air the initial fluorescence yield on illumination with light (less than 690 nm wavelength) is relatively low, but rises rapidly to a higher level. Illumination by continuous long-wavelength light (measuring fluorescence with weak or intermittent short-wave light) causes fluorescence to decline towards the lower level. The presence of oxidizing agents such as ferricyanide also depresses the level of fluorescence. As mentioned in Chapter 6, this behaviour is explained by supposing that fluorescence is quenched by a quencher, Q, that the quencher is reduced by system II restoring the fluorescence, and that Q is reoxidized by system I or by oxidizing agents.

Malkin and Kok showed that Q was not reduced until after the reduction of Hill reagents such as ferricyanide, and that the quantum yield of the reduction of Q was unity. This measurement is explained in the review by Kok and Cheniae (1967). Malkin showed further that there was an endogenous pool of oxidant causing a marked inflection in the time-curve of fluorescence (Figure 8.3) and termed this pool P. Knowing the number of quanta required to fully reduce the pool (Q + P) he could calculate the number of oxidizing equivalents present, which came out at between $\frac{1}{8}$ and $\frac{1}{30}$ of the number of molecules of chlorophyll, in different samples. At low temperatures (77°K) Q, but not P, was still reduced by system II; the quantity of P could thus be estimated, and was found to be approximately equal to Q: $\frac{1}{16}$ to $\frac{1}{60}$ equiv./mole chlorophyll.

Joliot had meanwhile studied the quantities of oxygen released by flashes of light of various durations. He found that there was a small pool, E, which was sufficient to allow molecules of oxygen to be produced by the shortest flashes, and two larger pools, A_1 and A_2, which allowed a larger quantity to be evolved from longer flashes. The sizes of Joliot's pools A_1, A_2 were close to the sizes of Malkin's pools Q and P. The terminology is now complicated, since the symbol Q has been

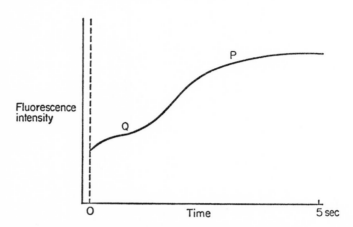

Figure 8.3. Time-course of fluorescence observed during illumination of isolated chloroplasts with short-wave light, after a period of darkness, or after an exposure to far-red light. The successive removal of two distinct pools of quenching material (Q and P) can be seen clearly. From Kok, B and Cheniae G. M. (1965). In D. R. Sanadi (Ed.) *Current Topics in Bioenergetics*, Vol. 1. Academic Press, New York, p. 1, with permission

almost universally used to denote the quencher (following Duysens and Sweers); the primary electron acceptor of reaction II is however E on the above scheme, Q (A_1) being a separate pool of oxidant, and E is now more likely to be the actual quencher.

We have referred in Chapter 6 to work by Cramer and Butler, who measured fluorescence of chloroplasts during titration of the suspending medium between -400 mV and $+100$ mV. They found two plateaus (see Figure 8.4) indicating redox materials with E_0' (pH 7) of -20 to -35 mV and -270 to -320 mV. The less negative potential may be thought of as an improved estimate of the standard potential of Q, obtained by Kok and Owens, of $+180$ mV. (The more negative potential has no place on the zig-zag scheme, at least near system II and is not accounted for.) A potential of

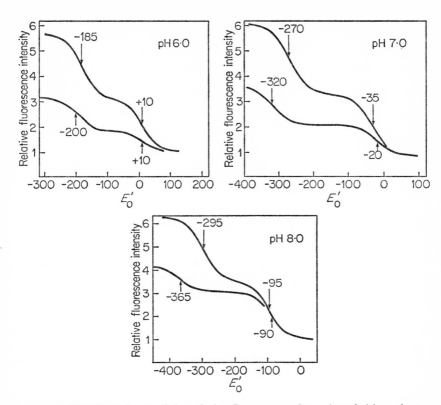

Figure 8.4. The dependence of the relative fluorescence intensity of chloroplasts on the redox potential of the medium, varied first by titration with $K_3Fe(CN)_6$ (left to right), secondly with $Na_2S_4O_6$ (right to left). The positions of maximum slope indicate the standard potentials of quenching materials in the thylakoid. No further changes were observed up to +500 mV. The variation of E_0' with pH is of the order of 60 mV, i.e. both quenchers dissociate one H^+ per electron during oxidation. From W. A. Cramer and W. L. Butler (1969). *Biochim. Biophys. Acta* **172**, 503, Figs. 1, 2 and 3, with permission

say −27 mV is convenient to associate with the potentials of the plasto-quinones measured by Carrier (1966): in ethanol, E_0' was +55 mV for PQ-C and +115 mV for PQ-A. The abundance of PQ-A in chloroplasts is of the order of $\frac{1}{7}$ mole per mole of chlorophyll, but a variable part of this, often half, is present in the osmiophilic globules which may take no part in electron transport. The corresponding figure for PQ-C is only $\frac{1}{30}$ of the chlorophyll. We have two plastoquinones, to identify with E, $A_1(Q)$ and $A_2(P)$. We may suppose that PQ-C is oxidized in the natural sequence by PQ-A, in accordance with the order of their standard potentials. (It is possible that they account for the 'A' pools. It is also possible that PQ-C is the E pool, and PQ-A then either A_1 or A_2.) However from the kinetic data of Malkin and Joliot there is reason to believe that E and A were in equilibrium, that is, that they were of similar standard potentials. The gap of 60 mV between the standard potentials of the quinones is a little large.

Returning to the spectroscopic data obtained by Witt's group, we recall that successive changes at 320 and 263 nm were ascribed to the formation

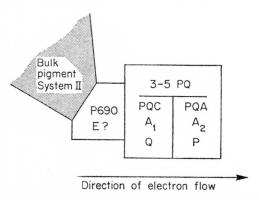

Direction of electron flow

Figure 8.5. A scheme for the site X_{II} relating and identifying (tentatively) various intermediates. The shaded area refers to the 'funnel' picture of the photosynthetic unit

first of a semiquinone of (a) plastoquinone, and secondly the disproportionation of the semiquinone to quinone and quinol. The formation of the semiquinone was extremely rapid (less than 2×10^{-8} s) which implied close electronic contact between PQ and the reaction centre, there being no time for the diffusion of a molecule. A similar electronic contact must be assumed for the reduction of $Q(A_1)$ by E. But the electron donor to PQ was considered by Witt to be the chlorophyll P690. We could therefore identify E with P690, $A_1(Q)$ with PQ-C and A_2 with PQ-A, as in Figure 8.5.

Both cytochromes b_6 and 559 have been claimed to be reduced by system II. Discussion of b_6 is held over to section 8.24. Cramer and Butler claimed to have observed the reduction of cytochrome 559 by system II, but Avron and Chance disputed this, and we have seen that Knaff and Arnon claim that cytochrome 559 is in fact oxidized by system II. Levine and Gorman have shown that mutants of *Chlamydomonas* lacking cytochrome 559 are inhibited in electron transport at a site close to system II, but on the one hand this is not inconsistent with cytochrome 559 being at Y_{II} instead of X_{II}, and on the other it is possible that the 559-component was not the high-potential cytochrome 559 but either a form of b_6 (known to split into 559 and 563 nm components at low temperatures) or yet a third form, perhaps like cytochrome b_3. At the present time it does not seem wise to attempt to place the b cytochromes in explicit sequences at either Y_{II} or X_{II}, but we should bear in mind that there is here data which cannot be accommodated under a simple scheme consisting of unique sequences at both these sites.

8.23 *Current schemes of cyclic and non-cyclic electron transport in thylakoids*

Figure 8.6 sketches the zig-zag model with the sequences of sections 8.21 and 8.22 marked on it. This model has been adapted to show cyclic electron

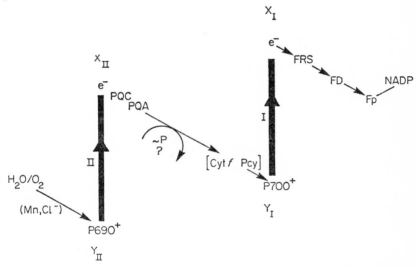

Figure 8.6. 'Zig-zag' scheme of non-cyclic electron transport summarizing section 8.21 and 8.22 (see text).

transport by indicating a path from the X_I region to a site before the phosphorylation step (Figure 8.7). Arnon has shown that of the naturally occurring chloroplast components, only ferredoxin acts as a catalyst for cyclic phosphorylation. This permits the use of Forti's (1966) discovery of a link (*in vitro*) between NADP-reductase and cytochrome f. Furthermore, if the pathway were according to Figure 8.8 with a phosphorylation step separate from that of the non-cyclic path, it would explain Arnon's observation that the two phosphorylation sites had differing sensitivities to the inhibitor desaspidin. Cytochrome b_6 has a potential close to that of PQ-C over the physiological

P. 130

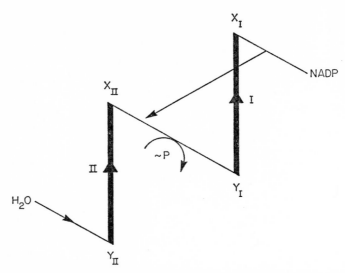

Figure 8.7. Modification of Fig. 8.6 to show the principle of cyclic flow around System I.

pH range, and could act here, as it occurs both in system I and II particles. On this basis the loop might make use of cytochrome b_6, as both Hind and Olson (1966) and Witt and coworkers (1969) have shown fast reduction of b_6 by system I light.

Hind and Olson (1967) found that cytochrome b_6 was reduced by both systems I and II. Such changes were also noticed by Witt's group (see

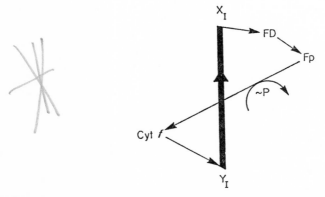

Figure 8.8. A more explicit version of Fig. 8.7, making use of Forti's NADP-reductase–cytochrome f complex, and suggesting a phosphorylation site off the non-cyclic pathway.

Rumberg, 1966) who concluded that b_6 was reduced by PQ in parallel with the reduction of cytochrome f. The same scheme was adopted by Hind and Olson (1967) (see Figure 8.9) who regarded plastocyanin as a possible sink for the electrons of reduced cytochrome b_6. In this, they were anticipated by Hill (1965) who had produced a very similar solution (Figure 8.10).

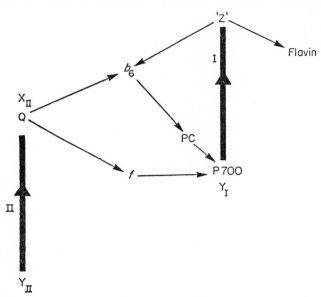

Figure 8.9. Scheme of G. Hind and J. M. Olson (1967) based on the observed action of inhibitors. From *Energy Conversion by the Photosynthetic Apparatus*, Brookhaven Symposium No. 19, p. 188, with permission

Arnon however chose to separate the two systems completely. Figure 8.11a is his (1965) model for system I, with the pathway for reduction of NADP by the ascorbate–indophenol donor included (Arnon, Tsujimoto and McSwain, 1965). This leaves the water–NADP pathway to be run by system II alone as a one-quantum process (Figure 8.11b). This part of his model appears somewhat weak, as the enhancement data (see section 6.1), the requirement for plastocyanin in the water–NADP pathway, and the close energy balance make a one-quantum process unlikely, besides which the requirement for a phosphorylation step at the Y_{II} site adds a further complication to an already difficult problem. Much of the objection may however be met by a new concept (Knaff and Arnon, 1969), referred to on page 138, in which system II contains two light reactions connected by an electron-transport pathway involving cytochrome 559. This is illustrated in Figure 8.11c.

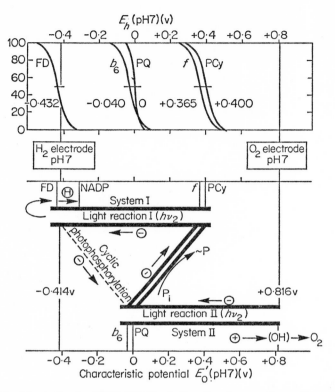

Figure 8.10. A hypothesis of Hill (1965) relating the zig-zag scheme (lower diagram) to the ranges of redox potential of the carriers (upper diagram). Allowance is made for separate pathways between X_{II} and Y_I. From P. N. Campbell and G. D. Greville (Eds.), *Essays in Biochemistry*, Vol. 1, Academic Press, London, p. 121, with permission

A theory has been proposed by Franck and Rosenberg in which there is only one reaction centre, but it receives both singlet and triplet excitation collected by the pigment systems. This model can explain the enhancement phenomena. The reaction centre is supposed to act twice on a redox carrier such as cytochrome f (which represents X_{II} and Y_I) and generates the oxidant and reductant (Y_{II} and X_I) which produce oxygen and reduce NADP. The existence of the other redox materials with intermediate potentials is not in accord with the theory, and the fractionation of the thylakoid into I and II particles is a further difficulty.

Warburg's hypothesis, in which light splits activated carbon dioxide directly to carbohydrate and oxygen, three-quarters of which then recombine

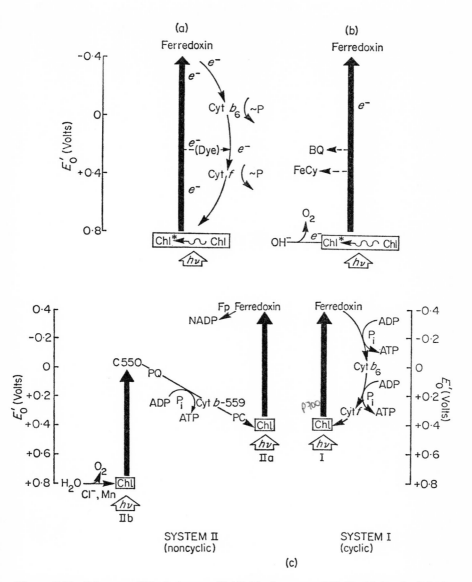

Figure 8.11. Schemes of D. I. Arnon, H. Y. Tsujimoto and B. D. McSwain (1965) for (a) cyclic and (b) non-cyclic electron transport (from *Nature* **207**, 1367, with permission) and (c) of D. B. Knaff and D. I. Arnon (1969) for a three-light-reaction scheme in which System II is a zig-zag of reactions IIb and IIa, and System I as before is reserved for cyclic electron flow. From *Proc. Natl. Acad. Sci.* **64**, 715, with permission

147

as in respiration producing the energy for the activation of the carbon dioxide, demands the complete reversal of the foregoing concepts of electron transport, in favour of pathways not yet discovered. The hypothesis is based experimentally on an apparent quantum requirement of four per carbon dioxide molecule fixed, on the (agreed) requirement for carbon dioxide for maximum Hill reaction activity, and of an apparent store of loosely bound carbon dioxide in algae. A summary of Warburg's position can be found in Warburg and coworkers (1969). One might leave this scheme as an exercise in model criticism, or acknowledge, following Hill (1965), that it has played a most useful part in stimulating interest, experiment and discussion in this field.

It is hard to escape the conclusion from this survey that the zig-zag model presents more satisfactory explanations for a greater proportion of the observations than any of its rivals.

8.3 Sequences in some photosynthetic bacteria

The bacterial system has some close analogies with that of the chloroplast. There are at least two spectral forms of the main pigment, one the light-harvesting pigment and the other a reaction-centre-specific form, usually but not always at a longer wavelength. Part of the reaction-centre pigment is photoreactive, and shows reversible bleaching by light. At pH 7·5 the standard potential of P890 in *Chromatium* is $+0·49$ V, and $+0·44$ V at pH 7·4 in *R. rubrum*. Parson (1968) has shown that the oxidation of P890 in *Chromatium* chromatophores is extremely rapid (less than 0·5 μs) and is followed by the oxidation of cytochrome 555 (C555) with reduction of P890 at 2 μs. The oxidation of C555 is a thermal process. There is a second cytochrome, C552, which is oxidized in a non-thermal reaction with a half-time of about 2 ms. Cusanowitch and coworkers (1968) have suggested that there is a second light reaction operating on C552, and Morita,* (1968) has shown that the action spectra for the two oxidation reactions differ: in particular, that the bacteriochlorophyll species B800 and B850 contribute relatively less than B890 to the C555 oxidation than to the C552 oxidation. (In this notation B890 differs from P890 in that it is not reversibly photobleached. It is presumed to be a light-harvesting pigment close to, or part of the reaction centre.) Hind and Olson (1968), using the symbol P'890 for the second centre oxidizing C552, have produced the scheme shown in Figure 8.12 for the photosynthetic electron transport pathways in *Chromatium*. This scheme would explain how NAD may be reduced by non-cyclic electron transport via ferredoxin, or by reverse electron transport. It rationalizes the observed ability of thiosulphate

* Morita in fact prefers a three-light-reaction hypothesis.

to reduce C552, and thence to photoreduce ferredoxin, and of succinate to reduce NAD without any requirement for ferredoxin. The ferredoxin-dependent reduction of NAD has been shown to proceed by means of a soluble reductase (Weaver and coworkers, 1965). The scheme is not certain however, and a short list of observations which are not readily explained by it is given in the review of Hind and Olson (1968). This review also gives a summary of cytochromes and their changes observed in bacteria other than *Chromatium*.

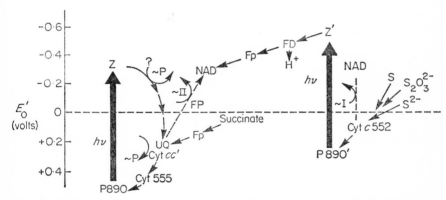

Figure 8.12. Scheme presented by G. Hind and J. M. Olson (1968) to account for observations on Chromatium, by means of a two-light-reaction model. The scale indicates the standard potentials at pH 7·5. ∼I represents a high-energy intermediate before ∼P (high-energy phosphate). UQ: ubiquinone; Z: hypothetical acceptor (corresponding to X in other figures). The dashed lines represent proposed 'reverse electron transport' pathways. From *Ann. Rev. Plt. Physiol.*, **19**, 249, with permission

The scheme of Hind and Olson (Figure 8.12) uses the symbols Z, Z′ for the primary acceptors (X in the usage of this text). There has been considerable discussion whether these acceptors may be chlorophyll molecules. Krasnowskii has demonstrated reduction of chlorophylls in organic solvents but on the other hand, the chloroplast acceptor may well be plastoquinone (PQ) (see section 8.22) and quinones are available for consideration in this role in bacteria. Ubiquinone (UQ) occurs throughout the purple bacteria, and naphthoquinones are found in green bacteria. The standard potentials of UQ and naphthoquinones of the K-vitamin series at 25° and pH 7·5 (the pH for which Figure 8.12 is drawn) are +0·092 and −0·036 V. Potentials of this order of magnitude have been found by titrating chromatophores from various species with redox buffers until the bleaching (oxidation) of P890 was diminished by 50%. However the direct observation of quinone changes by spectrophotometry at 270 nm (UQ) have not supported the concept; Figure 8.12 avoids this by placing UQ after Z in a cyclic pathway. This is supported by the observed inhibition of the cyclic electron transport system when the quinones are extracted.

References

Amesz, J. and D. C. Fork (1967). *Photochem. Photobiol.*, **6**, 903.

Anderson, J. M., D. C. Fork and J. Amesz (1966). *Biochem. Biophys. Res. Commun.*, **23**, 874.

Arnon, D. I. (1967). In T. W. Goodwin (Ed.), *Biochemistry of Chloroplasts*, Vol. 2, Academic Press, London, p. 461.

Arnon, D. I., H. Y. Tsujimoto and B. D. McSwain (1965). *Nature*, **207**, 1367.

Avron, M. and B. Chance (1966). In J. B. Thomas and J. C. Goedheer (Eds.), *Currents in Photosynthesis*, Donker, Rotterdam, p. 455.

Bendall, D. S. (1968), *Biochem. J.* **109**, 46P.

Bertsch, W. F. (1969). In H. Metzner (Ed.), *Progress in Photosynthesis Research*, Vol. 2, Institut für Chemische Pflanzenphysiologie, Tubingen, p. 996.

Bothe, H. (1969). In H. Metzner (Ed.), *Progress in Photosynthesis Research*, Vol. 3, Institut für Chemische Pflanzenphysiologie, Tubingen, p. 1483.

Carrier, J. M. (1967). In T. W. Goodwin (Ed.), *Biochemistry of Chloroplasts*, Vol. 2, Academic Press, London, p. 551.

Cramer, W. A. and W. L. Butler (1969). *Biochim. Biophys. Acta*, **172**, 503.

Cusanowitch, M. A., R. G. Bartsch and M. D. Kamen (1968). *Biochim. Biophys. Acta*, **153**, 397.

Davenport, H. E. and R. Hill (1952). *Proc. Roy. Soc. B*, **139**, 327.

Emerson, R. and W. Arnold (1932), *J. Gen. Physiol.*, **16**, 191.

Forti, G. and G. Zanetti (1967). In T. W. Goodwin (Ed.), *Biochemistry of Chloroplasts*, Vol. 2, Academic Press, London, p. 523.

Fujita, Y. and J. Myers (1967). *Arch. Biochem. Biophys.*, **119**, 8.

Gorman, D. S. and R. P. Levine (1966). *Plant Physiol.*, **41**, 1648.

Henninger, M. D. and F. L. Crane (1967). *J. Biol. Chem.*, **242**, 1155.

Hill, R., (1965). In P. N. Campbell and G. D. Greville (Eds.), *Essays in Biochemistry*, Vol. 1. Academic Press, London, p. 121.

Hind, G. and J. M. Olson (1967). *In Energy Conversion by the Photosynthetic Apparatus*, Brookhaven Symposium No. 19. p. 188.

Hind, G. and J. M. Olson (1968). *Ann. Rev. Plant Physiol.*, **19**, 249.

Izawa, S., R. L. Heath and G. Hind (1969). *Biochim. Biophys. Acta.*, **180**, 388.

Joliot, P., G. Barbieri and R. Chabaud (1969). *Photochem. Photobiol.*, **10**, 309.

Keister, D. L., A. San Pietro and F. E. Stolzenbach (1960). *Arch. Biochem. Biophys.*, **98**, 235.

Klingenberg, M., A. Mueller, P. Schmidt-Mende and H. T. Witt (1962). *Nature*, **194**, 379.

Knaff, D. B. and D. I. Arnon (1969a). *Proc. Natl. Acad. Sci. U.S.*, **63**, 956.

Knaff, D. B. and D. I. Arnon (1969b). *Proc. Natl. Acad. Sci. U.S.*, **64**, 715.

Kok, B. and G. M. Cheniae (1966). In D. R. Sanadi (Ed.), *Current Topics in Bioenergetics*, Vol. 1, Academic Press, New York, p. 1.

Kok, B., B. Forbush and M. McGloin (1970). *Photochem. Photobiol.*, **11**, 457.

Levine, R. P. (1969). In H. Metzner (Ed.), *Progress in Photosynthesis Research*, Vol. 2, Institut für Chemische Pflanzenphysiologie, Tubingen, p. 971.

Lundegardh, H. (1962). *Physiol. Plantarum*, **15**, 390.

Morita, S. (1968). In K. Shibata and others (Eds.), *Comparative Biochemistry and Biophysics of Photosynthesis*, University of Tokyo Press, Tokyo, p. 133.

Parson, W. W. (1968). *Biochim. Biophys. Acta*, **153**, 248.

Rumberg, B. (1966). In J. B. Thomas and J. C. Goedheer (Eds.), *Currents in Photosynthesis*, Donker, Rotterdam, p. 375.

Shin, M., K. Tagawa and D. I. Arnon (1963). *Biochem. Z.*, **338**, 84.

Warburg, O., G. Krippahl and A. Lehman (1969). *Amer. J. Bot.*, **59**, 961.

Weaver, P., K. Tinker and R. C. Valentine (1965). *Biochem. Biophys. Res. Commun.*, **21**, 195.

Witt, H. T., B. Suerra and J. Vater (1966). In J. B. Thomas and J. C. Goedheer, *Currents in Photosynthesis*, Donker, Rotterdam, p. 273.

Witt, H. T., B. Rumberg and W. Junge (1969). In H. Staudinger and B. Hess (Eds.), *Biochemie des Sauerstoffs*, Mosbach Symposium 1968, Springer, Berlin. p. 262.

CHAPTER 9

Phosphorylation

From an early date, ATP was known to be formed from ADP and inorganic phosphate during oxidation of metabolites by the mitochondrion. ATP was found to be of great importance in the energy relations of the living cell, required for some obvious biochemical syntheses, and the contraction of actomyosin in muscle fibrils. Nearly all energy-consuming processes, such as the operation of flagella, contractile vacuoles, transport of many materials across the cell membrane, nervous conduction and luminescence of fireflies—in short, conversions of chemical energy to physical work—were inhibited by reagents such as dinitrophenol or cyanide which prevented the formation of ATP. However the mechanism by which ATP is formed has proved one of the most obscure biochemical problems yet encountered. At the present time there are two main hypotheses concerning the means of ATP production—termed *phosphorylation* (of ADP, understood)—both of which postulate for the crucial phases unknown intermediates denoted by symbols X, I and so on, for which very little chemical characterization has been forthcoming.

ATP was also, and later, found to be a product of chromatophores and chloroplasts in light, from the researches of Frenkel and of Arnon, and this phosphorylation has been shown to be of fundamental importance for the process of photosynthesis. Fortunately, although photosynthetic phosphorylation is as obscure as that in the mitochondria, it can nevertheless be seen that the two problems are much alike, and we can profit by considering them together in this chapter.

In the aerobic respiration of glucose to carbon dioxide and water by the pathways of glycolysis and the citric acid cycle, 40 moles of ATP per mole of glucose are formed (two are consumed), and of these 40, 34 are formed by the process of oxidation of reduced coenzymes by oxygen in the mitochondrion. This is *oxidative phosphorylation*. The remaining 6 ATP are formed by enzymes that react with metabolic substrates, and this is termed *substrate level* phosphorylation. The two processes together are sometimes termed oxidative phosphorylation, and the term *respiratory chain level phosphorylation* may be used to indicate the mitochondrial process. At substrate level the oxygen atom eliminated by the condensation from ADP and P_i is incorporated into the substrate, but at respiratory chain level it is eliminated as water. *Photophosphorylation* is the formation of ATP by chloroplasts or chromatophores given ADP, P_i and magnesium ions, under illumination. Since no metabolites are immediately involved, the process has more in common with respiratory chain level than with substrate level oxidative phosphorylation. The methods and concepts used in investigating photophosphorylation are in the main those derived from studies on the mitochondrion. Therefore it will be appropriate to outline these methods and introduce the terminology by a survey of oxidative phosphorylation.

9.1 Oxidative phosphorylation

Figure 4.4 sets out on a redox-potential diagram a sequence of carriers which probably represent the mitochondrial electron transport chain. Passage of two electrons (reducing one half-molecule of oxygen) from NADH to one half-molecule of oxygen results in the formation of three molecules of ATP. This 'P/O' or 'P/2e$^-$' ratio on the other hand is only two when succinate is the electron donor. This is taken to indicate that there are three phosphorylation sites on the NADH pathway, of which two are common to the succinate pathway. Further location of these sites is arrived at by three lines of reasoning. First, inspection of Figure 4.4 shows wide potential gaps between adjacent redox carriers in three or possibly four cases: NADH–FAD, FAD–ubiquinone, ubiquinone–cytochrome c_1, cytochrome a–oxygen. Secondly, by applying reagents which inject or remove electrons from more or less specific sites in the chain, one can measure the P/2e$^-$ ratio in the part of the chain thus isolated. The results from such experiments are illustrated in Figure 9.1. Thirdly, the *crossover* method of Chance makes use of the inhibition of electron transport in mitochondria supplied with P_i but deficient in ADP (state 4). When ADP is added, electron transport accelerates and the steady-state redox level of the redox carriers is altered, as observed spectroscopically. Redox carriers to the oxygen side of the substrate side become more oxidized. Thus the 'crossover point' indicates the rate-limiting phosphorylation site. The entire observation, repeated in the presence of low and high concentrations of an inhibitor such as azide shows the three sites indicated in Figure 9.2.

It should be noted that at least part of the electron transport pathway of mitochondria is reversible when coupled to phosphorylation. Given ATP, aNAD$^+$ and succinate, mitochondria can form NADH (and fumarate), splitting the ATP which provides the energy necessary to drive the electrons to the more negative potential. This process is known as *reverse electron*

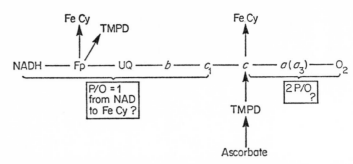

Figure 9.1. Diagram of the mitochondrial electron transport pathway, showing its interactions with artificial electron acceptors and donors, and the P/O ratios obtained. However note that the value of 1 for the section NAD–FeCy is to be considered as an approximation to 2, and the value of 2 for the latter part is considered to be brought about by a contribution from the section NAD–Fp–TMPD, even in the presence of the inhibitor amytal. FeCy: ferricyanide; TMPD: tetramethylphenylenediamine. After M. Klingenberg (1968). In T. P. Singer (Ed.), *Biological Oxidation*, Interscience, New York, p. 3, with permission

transport. In the same way, an *energy-linked transhydrogenase* can be demonstrated which reduces $NADP^+$ using ATP and NADH. Both these activities can be driven by the energy released by the oxidation of succinate by oxygen in the remainder of the respiratory chain, the energy being transferred in some form other than ATP (e.g. $X \sim I$). Both processes are known in the chromatophores of photosynthetic bacteria, using the energy of light, ATP or pyrophosphate.

Evidence regarding the mechanism of ATP formation is derived from various kinds of experiment. First there is the study of *uncouplers*, which accelerate electron transport in state 4 mitochondria, and reduce the P/O ratio in phosphorylating (state 3) mitochondria. Typical uncouplers are arsenate and 2,4-dinitrophenol. Secondly, there are inhibitors such as oligomycin which prevent the formation of ATP without accelerating electron transport in state 4 mitochondria; in state 3, electron transport is slowed down, so that the P/O ratio is not reduced. In the presence of both

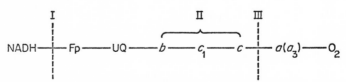

Figure 9.2. As in Fig. 9.1, showing phosphorylation sites deduced from the 'crossover theorem' (see text).

oligomycin and dinitrophenol only the effects of dinitrophenol are observed. Thirdly, *ion movements* can be demonstrated in respiring mitochondria, and these take place at the expense of the formation of ATP. The mitochondrion can take up approximately two Ca^{2+} ions (depending on the conditions) per half-molecule of oxygen reduced; $6H^+$ ions are extruded at the same time when the reductant is NADH, and $4H^+$ with succinate. Strontium (Sr^{2+}) and Mn^{2+} ions behave similarly. The uncoupler valinomycin permits the uptake of K^+ ions on the same scale. Oligomycin does not inhibit these ion-movements when they are driven by respiration but it does so when the same movements are driven by exogenous ATP in non-respiring mitochondria.

A fourth line of attack on this problem derives from the preparation of *coupling factors* from the small stalked spheres that can be seen on the cristae of mitochondria and removed by ultrasonic vibration with ethylenediamine tetraacetate. These stalked spheres were illustrated in Plate 9 (facing p. 84). From these particles proteins have been isolated and termed coupling factors since they restore the phosphorylating potency to inactive material (see Racker and Conover 1963). The factor F_1, a protein of molecular weight 280 000, has Mg^{2+}-dependent ATPase activity. ATPases constitute a fifth line of attack in themselves, since their activity is likely to be a deranged form of an ATP-utilizing system, or the ATP-synthesizing system itself. Here it is noted that when the F_1 factor is combined with the F_0 factor (obtained from the mitochondrial membrane), the ATPase becomes sensitive to oligomycin. These stalked spheres are hence likely to be intimately connected with phosphorylation.

As remarked earlier, there are two principal types of theory which attempt to rationalize these results, the 'chemical intermediate' and the 'chemiosmotic' formulations. Both of these theories recognize that the energy released by the oxidoreduction reaction step must be conserved in at least two forms before either P_i or ADP is combined, and ATP formed. The difference lies in the postulated form that these energy stores take. The chemical intermediate theory looks for identifiable complexes of each of three redox carriers C_1, C_2 and C_3 with intermediates I_1, I_2 and I_3; these complexes CI are high energy complexes, and are often represented by a tilde ($\sim$) thus: $C \sim I$. The chemical intermediate theory continues by supposing that C is displaced by another unknown reactant, X, giving a second high energy complex, $X \sim I$. I can then be displaced by P_i (or possibly ADP) giving $X \sim P$, and finally X is displaced by ADP giving ATP (see Figure 4.12, p. 66). X and possibly I may be common to all three phosphorylation sites, or there may be site-specific intermediates X_1, X_2, X_3 and so on.

The 'chemiosmotic theory' put forward by Mitchell (see Mitchell, 1966), regards the first energy store not as a chemical complex but as an electro-chemical gradient across the membrane. In its simplest form this could be a pH gradient formed by pumping H^+ ions out of the mitochondrion during electron transport. The H^+ ions could then be exchanged with, say, K^+ ions so as to introduce an electric potential difference, the combined effect of the chemical and electric gradients being the 'protonmotive force'. Mitchell points out that this could be achieved by arranging the chain in the membrane according to Figure 9.3. Some of the redox carriers of the chain are of the A/AH_2 type: when these are reduced by a carrier of the A/A^- type a proton must be taken up from the aqueous phase, and similarly, when AH_2 is oxidized by an electron carrier, a proton is released. Mitchell suggested that

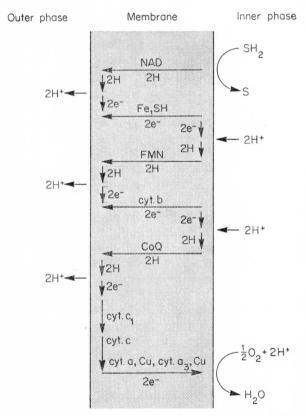

Figure 9.3. An arrangement of the mitochondrial electron transport chain, drawn to demonstrate a means for the forced translocation of H^+ ions across the membrane. From P. Mitchell (1966). *Biol. Revs.* **41**, 445, with permission

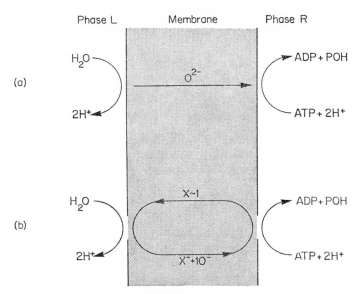

Figure 9.4. Proton-translocating reversible ATPase system. A: principle; B: operation with a hypothetical carrier, the anhydride $X \sim I$. From P. Mitchell (1966). *Biol. Revs.* **41**, 445, with permission

these processes take place at opposite sides of the membrane so that protons are pumped across. The next energy store in this hypothesis is a complex $X \sim I$ which shuttles across the membrane reacting on one side with ADP and P_i giving ATP:

$$X \sim I + ADP + P_i \xrightarrow{\text{alkaline pH}} X^- + IO^- + ATP + 2H^+$$

$$\text{or } XH + IO^- + ATP + 2H^+$$

$X \sim I$ has the properties of an anhydride (or possibly an ester). At the other side of the membrane X^- and IO^- condense to reform $X \sim I$ taking up protons:

$$X^- + IO^- + 2H^+ \xrightarrow{\text{acid pH}} X \sim I + H_2O$$

Strictly, XI is only a high-energy compound in the alkaline environment. The two reactions together constitute either an ATP-driven H^+ ion pump, acting in reverse, or an H^+-driven ATP synthetase, shown in Figure 9.4. The magnitude of the gradient necessary is at least 0.21 V (electric potential) or 3.5 pH units (chemical driving force), or a combination of the two, depending on the conditions; for a two-electron, or two-proton, transfer, this value converts to 9·7 kcal/mole, which is an approximate minimum for the formation of ATP under physiological conditions.

Discrimination between the two theories is not at present possible. So far it can be said that the chemiosmotic theory is at an advantage in that only one unknown complex is postulated, ion-movements are rationalized, and the action of weak acid uncoupling agents such as dinitrophenol is explained on the basis that they transport H^+ across the membrane destroying the pH gradient. Attempts to isolate any of the complexes of either theory have been unsuccessful, and the only evidence for the existence of a complex is the observation by Chance that in tightly-coupled mitochondria a spectral band of a cytochrome appears at 555 nm, vanishing with uncoupling agents. However, such a band could be due as much to a pH shift as to a high-energy complex.

The chemical-intermediate hypothesis demands that at each phosphorylation site in the electron transport chain there should be a redox carrier $C_{1,2,3}$ which forms a high-energy compound with the intermediate I. However a variant of the hypothesis allows for the complex $C \sim I$ to be a high-energy conformational state of a protein associated with C, as if C and I were groups on the same protein molecule. Apart from the observations of new spectral peaks alone, the conformational hypothesis has not so far suggested any discriminatory experiment. It has however been claimed that the contractile proteins obtained from mitochondria and chloroplasts operate in this role.

It is essential that there should be a 'sidedness' in phosphorylating membranes if the chemiosmotic hypothesis is correct. In the mitochondrion, ion-movements suggest that if the chemiosmotic theory holds then the inside of the mitochondrion is rendered alkaline by electron transport, hydrogen ions being expelled. Work done on vesicles prepared from mitochondria with digitonin appears to contradict this, but electron microscopy shows that the stalked spheres ('coupling factors') are now on the outside of the vesicles, in other words the membrane is inside-out. This shall be borne in mind when discussing the chloroplast.

9.2 Photosynthetic phosphorylation

Phosphorylation sites. The zig-zag scheme (Figures 4.10 or 8.6) may be inspected for potential gaps. Clearly there is a gap between PQ and cytochrome f of some 0·24 V. Other sites, at potentials more positive than $+0·8$ V or more reducing than $-0·4$ V, cannot be ruled out. In Arnon's scheme (see section 8.3, Figure 8.11a), there are two possible sites for cyclic phosphorylation, one between ferredoxin and cytochrome b_6, and one between cytochromes b_6 and f. The non-cyclic scheme (Figure 8.11b) requires phosphorylation sites at potentials outside the range $-0·4$ V to 0·8 V.

There is a basic difficulty that the $P/2e^-$ ratio for the non-cyclic process is uncertain; the figure usually measured is about 1, but since the rate of electron transport in the absence of ADP is relatively high (compared with the state 4 mitochondrion), the value could be calculated in terms of the *additional* electron transport when phosphorylation takes place, and a value of 2 is then obtained. In cyclic photophosphorylation there is no way yet of measuring $P/2e^-$ ratios; however observation of the effects of uncouplers can give information as to the number of sites. Thus in the chloroplast, Arnon (1967) has shown that the inhibitor *desaspidin* has different critical concentrations for the uncoupling of cyclic and non-cyclic phosphorylations. It may be that this result is due to an effect of oxygen on desaspidin, but it may alternatively indicate that the phosphorylation sites are not common to the two pathways.

Chance has applied the crossover technique to chloroplasts (Avron and Chance, 1966), from which it appears that there is a phosphorylation site before the reduction of cytochrome f in non-cyclic electron transport. The nature of the reductant was not established (referred to as a 'pool'), but might be assumed to be PQ.

There may also be more than one phosphorylation site in the redox chain of bacterial photosynthethesis. Thus the uncoupler valinomycin inhibits cyclic photophosphorylation to some extent, but phosphorylation is not valinomycin-sensitive if part of the pathway is by-passed by phenazine methosulphate (see Baltscheffsky and von Stedinck, 1966). Uncoupling agents have also been used to show that oxidative phosphorylation and photophosphorylation in *Rhodospirillum rubrum* (in chemotrophic and phototrophic modes of growth respectively) occur at different sites.

Polarity. It is interesting from the standpoint of the chemiosmotic hypothesis that photophosphorylation seems to be of opposite polarity to oxidative phosphorylation. Thus the chloroplast and chromatophore both accumulate hydrogen ions, whereas the mitochondrion extrudes them; thus the inside of thylakoids becomes more acid than the outside, whereas mitochondria, at least in the presence of calcium ions, become alkaline inside. The weak-base uncouplers of the chloroplast, the primary amines and ammonium salts, have little effect on the mitochondrion, which is the more powerfully affected by the weak-acid types. Lastly, the 9 nm stalked spheres occur on the outside of the photosynthetic membranes, but on the inside of the mitochondria.

Coupling factors. The 9 nm stalked spheres appear to contain as a principal component the factor F_1 (in mitochondria) or CF_1 (in chloroplasts)* (see

---

* It so happens that *chloroplast* factors have been named CF_1, CF_2, CF_c, etc, and that CF_1 is analogous to F_1 in mitochondria. Kagawa and Racker have however allotted the name CF_0 to a preparation derived from the mitochondrial factor F_0 by means of cholic acid: to that extent the nomenclature is inconsistent.

McCarty and Raker, 1967). A list of protein factors related to oxidative phosphorylation in mammalian mitochondria is given by Lardy and Ferguson (1969). F_1 is an ATPase, and CF_1 becomes an ATPase after a brief activation by trypsin. Several materials of this type, listed by Lardy and Ferguson as phosphoryl transferases, have been isolated from mitochondria, and may well represent the final enzyme in the pathway that produces ATP in respiration. By analogy, CF_1 may be presumed to be the final enzyme involved in photophosphorylation. Although CF_1 has no ATPase activity before treatment with trypsin, neither does the chloroplast; ATPase activity can also be induced in chloroplasts by treatment with light or an acid–alkali transition (both of which are assumed to generate a high-energy state $\sim$) in the presence of reducing agents such as dithiothreitol, and CF_1 similarly develops ATPase activity given dithiothreitol and thylakoid material. Both F_1 and CF_1 restore phosphorylation to membrane material from which the stalked spheres have been removed.

Trypsin is believed to act on chloroplasts or CF_1 by destroying a controlling-protein, CF_2 (Livne and Racker, 1968). The effect does not take place however on a protein CF_0 believed to differ from CF_1 only by the complement of metallic cations (Lynn and Straub, 1969a). Lynn and Straub (1969b) regard photophosphorylation as driven by an enzyme having both protonation and oxido-reduction sites.

A chemiosmotic or chemical-intermediate mechanism? The data on photophosphorylation in chloroplasts and in bacterial chromatophores can be interpreted on the chemical intermediate theory, and the work on the effect of valinomycin referred to above (Baltscheffsky and von Stedinck 1966) was taken to indicate that one phosphorylation site led to the intermediate $X \sim I$ which in the presence of valinomycin was diverted to ion-transport, while the other site led to $X \sim P$ or $X \sim ADP$ via a different path. Even if the valinomycin-sensitive path $(X \sim I)$ can be interpreted on the chemiosmotic hypothesis, the existence of an insensitive path immediately raises doubt. Nevertheless, the same authors demonstrated that chromatophores accumulated protons from the medium, and the accumulation was inhibited by uncoupling agents such as desaspidin and 2-n-heptyl-4-hydroxyquinoline-N-oxide (HOQNO). Addition of PMS however restored the ability to phosphorylate, but did not restore the proton uptake. Clearly these results do not allow one to come to any conclusion at this stage in favour of either the chemical-intermediate or the chemiosmotic hypotheses.

Impressive evidence for the chemiosmotic theory in chloroplasts came from an attempt by Hind and Jagendorf (1963) to isolate high energy intermediates 'X_E' by illuminating chloroplasts in the absence of ADP and allowing phosphorylation to occur only in the dark. The two steps were operated at different pH values, and it was observed that transference from an acid medium to an alkaline one would result in some ATP formation, (Jagendorf and Neumann, 1965). In acid media, of pH 5 or 6, illumination of chloroplasts (which would probably be devoid of their envelopes and most of the stroma) showed a strong uptake of H^+ ions. There was therefore a strong inference that X_E was the same as the pH gradient. Also the quantity of X_E (270/2500 chlorophylls) was greater than that of any known redox carrier, including PQ. Mitchell (1966) pointed out that chloroplasts with intact membranes are less dependent on the external pH value; the significant parameter is the 'proton motive force' (PMF), of which pH is only one component.

The absorption change at 515 nm observed in chloroplasts by Witt (1968) takes place in less than 2×10^{-8} s, and is the fastest 'reaction' detected so far. It is non-thermal (that is, the rate does not depend on temperature), and therefore presumably entirely an electronic change. Uncoupling agents and phosphorylation accelerate the decay of the 515 nm change, and since uncouplers (such as gramicidin) that make the thylakoid membrane permeable to specific ions (such as K^+) are as effective as those which transport hydrogen ions, Witt regards the 515 nm difference as due to an electric field resulting from electron displacements within the thylakoid membrane. Such a potential difference while of small magnitude would amount to a severe electric field since the thylakoid is such a thin membrane; pigments are known to alter their absorption spectra slightly in a strong electric field; this is called the Stark effect. Witt has also been able to demonstrate that proton fluxes take place into the inner space of the thylakoid. These are driven by the electron transport, and operate by the mechanism of Figure 9.3, one proton being released from photoreaction I and one from photoreaction II in a short flash (less than 6×10^{-4} s). The PQ pool holds 10 electrons, and in a longer flash ($1 \cdot 5 \times 10^{-2}$ s) a proton influx correspondingly greater is observed. Finally if the thylakoid is allowed to carry out electron transport in steady light, the proton influx continues until the ion capacity of the thylakoid sac becomes saturated, with an influx 100 times greater still. The proton gradient was assumed to be contained by the thylakoid membrane as a whole, since gramicidin uncoupled by 50% at a concentration of one mole per 2×10^5 chlorophylls. Since 10^5 chlorophylls cover an area of 500×500 nm (the average size of a thylakoid), it appears that one gramicidin molecule is sufficient to perforate one thylakoid.

The same group (see Rumberg and Siggel, 1969) have shown that the electron transport between PQ and P700 is accelerated by phosphorylation and uncoupling agents. In the presence of the uncoupler gramicidin (at high concentration), the rate of the reaction depends on the pH of the outer solution, and it is assumed that the gramicidin, now making the membrane permeable to hydrogen ions, causes the internal pH to be equal to that of the outer phase. By this means the rate of the PQ—P700 reaction is obtained as a function of the (internal) pH. When electron transport is allowed to proceed by illuminating a chloroplast in the absence of uncoupler, and with no ZDP or P_i, the rate of reaction PQ—P700 declines rapidly from its initial rate. This is seen as an indication that the internal pH is falling, and the calculated value of the final internal pH is 5·1 when the external pH is 8·0. When phosphorylation is allowed to proceed the pH difference narrows to 2·7 units.

Before this pH difference can be related to the free energy change required for ATP synthesis, it is necessary to know what the value of the electrochemical potential is across the membrane. Witt's group have used the 515 nm change to measure this, on the assumption that it does indeed represent a Stark effect. Assuming that the initial Stark effect is due to the production of two positive–negative charge-pairs separated by a lipid layer 3 nm thick (dielectric constant assumed to be 2), in an area of 10^3 nm^2 (for one complete electron transport chain), they calculate the initial field as 50 mV by the condenser formula. Assuming further that the Stark effect is linear with the field, they calculate the increase in the field from the magnitude of the 515 nm change as it builds up with time. They conclude that the electrochemical field across the membrane has a steady-state value of some 100 mV, with a possible maximum of 200 mV. The total energy stored in a pH difference of 2·7 units and an electric field of 0·2 V is:

Electric: $F \times \Delta E$ J/mole $H^+ = \dfrac{96\ 500 \times 0·2}{4·184}$ cal/mole H^+

$= 4613$ cal/mole H^+

pH: $2·303\ RT \log \Delta pH = 1364 \times 2·7$

$= 3682$ cal/mole H^+ (25°)

Total: 8295 cal/mole H^+ at 25°, rounded to 16 kcal/$2H^+$

16 kcal per pair of hydrogen ions transferred should allow ATP formation to produce a high value of the equilibrium $[ATP]/[ADP][P_i]$.

While the experiments of Hind and Jagendorf and of Witt's group provide overwhelming evidence that energy can be stored as a chemiosmotic gradient across the thylakoid membrane, there is still the possibility that this gradient is built up secondarily at the expense of a primary pool representing $C \sim I$ or $X \sim I$, so that the chemical intermediate hypothesis may still be tenable. It is clear that the chemiosmotic theory of phosphorylation provides an attractive framework for arranging and discussing many of the observations of photosynthesis. It would however be premature to discount the chemical intermediate approach, which also can account for most of the data.

9.3 Mechanisms proposed for phosphorylation

ATPases and exchange reactions. Since ATP is such an important material, ATPases, which merely hydrolyse ATP to ADP and P_i, are unlikely to be of biological value to the cell, and must therefore probably be artefacts produced by the experimental conditions, representing enzymes which had previously been concerned with ATP utilization or synthesis. Thus the ATPases of nervous tissue in animals may be derived from the ATP-requiring sodium pump, and the coupling factors of Racker which have ATPase activity have in section 9.1 been considered as part of the ATP synthesizing machinery of both mitochondria and chloroplasts.

Unlike mitochondria, intact thylakoids do not show any ATPase activity in the dark. However a Ca^{2+}-dependent ATPase can be obtained from thylakoids; this enzyme is inhibited by the antibiotic Dio-9 (which blocks ATP formation in the chloroplast). This ATPase appears to be identical with the coupling factor. A second ATPase was observed by Avron, in chloroplast material, which required both calcium and (continuous) light. A third chloroplast ATPase requires magnesium ions, and needs to be induced by either a brief illumination or an acid–alkali transition in the presence of reducing agents such as dithiothreitol. This is termed the light-triggered ATPase. The effect of the triggering dies away rapidly unless ATP is present, when the ATPase activity persists unabated for many minutes.

For the sake of discussion, let us write a scheme for photophosphorylation, for example that of Figure 9.5.

In this scheme, the reversibility of the reactions would mean that if ATP were supplied it would give rise to a PMF across the thylakoid, and if the system were detached from the thylakoid, the energy would be dissipated and an ATPase action would result. This provides a rationalization of the ATPase of the coupling factor. Secondly, in the intact thylakoid, uncoupling agents which are believed to render the membrane permeable to H^+ ions and so collapse the PMF would bring about the hydrolysis of ATP in the same way. This is in fact observed in mitochondria, but not in chloroplasts. However uncoupling agents do have two effects in the chloroplast in this context: first they inhibit the triggering of the light-triggered ATPase, and secondly they stimulate it once it has begun. We could conclude from these results, and from the need for ATP to maintain the light-triggered ATPase, that ATP has a controlling action on the phosphorylating enzymes. 'Allosteric' effects of ATP on other enzymes are well known. A biological reason for such an effect is that plants must face regular periods of darkness, when they obtain their ATP from mitochondria. This 'switching off' of the chloroplast would save wasting ATP. In mitochondria activity may be continuous so that external ATP is always balanced by 'respiration pressure'. It does not therefore need to have such a switching-off mechanism.

Further information can be obtained from observing *exchange reactions*. This type of reaction is carried out under conditions where although no net

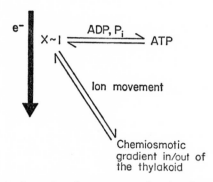

Figure 9.5. Diagram to show that the presence of a chemiosmotic gradient coupled to phosphorylation does not preclude a chemical-intermediate mechanism. X ∼ I may be either a real substance or a symbol for a pH gradient. The heavy arrow represents the electron transport pathway

changes take place in the quantity of the materials present, isotopic label is transferred from one substance to another. Thus if thylakoids, ATP and radioactive inorganic phosphate ($^{32}P_i$) are incubated together, some of the radioactivity may become incorporated into the ATP although the total quantity of ATP hardly changes. This is the ATP–P_i exchange, and it is significant that it needs to be triggered in the same way as the light-triggered ATPase. Reference to Figure 9.5 shows that an ATP–P_i exchange would be expected from the operation of the right-hand reaction of the sequence, that is the formation of X ∼ I, with the splitting of ATP to ADP and P_i. The P_i would mix with the radioactive P_i and the reverse reaction would re-synthesize ATP containing some ^{32}P. This locates the control-site at that particular reaction.

It has been noted however that there is no corresponding ATP–ADP exchange, not even under the light-triggering conditions. This implies that there is a difference in the way ADP and P_i react with the enzyme. If the enzyme acted with both ADP and P_i combining at two binding-sites simultaneously (this is dalled a 'concerted' mechanism), then both ATP–ADP and

ATP–P_i exchanges should occur. If on the other hand the enzyme initially formed a specific complex with only one reactant:

Either: $\qquad\qquad$ Enz. $+$ ATP $\rightleftharpoons$ Enz. ADP $+ P_i$ $\qquad\qquad$ (1)

$\quad$ *or:* $\qquad\qquad$ Enz. $+$ ATP $\rightleftharpoons$ Enz. P $\quad$ $+$ ADP $\qquad\qquad$ (2)

then in case (1) there should be an exchange of ATP with P_i but not with ADP, and in case (2) only the ATP–ADP exchange should be observed. Clearly case (1) fits the experimental findings.

Racker has disputed the above conclusion, on the ground that the ATP might indeed form a phospho-enzyme, but with the ADP still bound (under the conditions of the experiment). The observed ATP–P_i exchange would then be explained by a two stage system:

$$X \sim I \xleftrightarrow{\quad\quad} X \sim P \xleftrightarrow{\quad\quad} ATP \qquad (3)$$
$$\qquad\quad \downarrow \qquad\qquad\quad \downarrow$$
$$\qquad\quad P_i \qquad\qquad\quad ADP$$

This scheme (and scheme 2) also purports to explain the observed fact that the oxygen atom linking terminal (γ) and penultimate (β) phosphate groups in ATP comes entirely from the ADP. These considerations have generally led to the belief that phosphate is incorporated into the system before ADP. The scheme (3) above is relatively easy to relate to the chemical intermediate hypothesis. The chemiosmotic hypothesis further requires only that the chemical intermediate XI is preceded by an energy store in the form of the PMF.

There has been considerable discussion of the identity of X and I in terms of known chemical substances. The reader is referred to the reviews of Boyer (1968) and of Lardy and Ferguson (1969) for summaries of the various model systems suggested; although no phosphorylated derivatives have been discovered, the problem is perhaps easier once it is accepted that the phosphorylation intermediates X, I etc., need not be chemically combined with the electron transport carriers.

9.4 Summary

The formation of ATP by the electron transport pathways of the mitochondrion and of the chloroplast can be associated with certain redox reactions on the basis of (i) differences in the standard potentials exceeding say 150 mV, (ii) the 'crossover' theorem in which the redox levels of the carriers on either side of a phosphorylating site appear (spectroscopically) to converge as electron transport accelerates when ADP and P_i are added to a tightly coupled system, (iii) by injecting electrons into the pathway so as to isolate certain sections. The reader may feel that with the present uncertainty regarding the photosynthetic electron transport chains, particularly the cyclic one in chloroplasts, and chromatophores, these principles are of small practical use, but it can be claimed with some force that one phosphorylation site, at least, has been associated with the reduction of cytochrome f by PQ.

Both chloroplasts and mitochondria show uncoupling, by agents that break down or prevent the formation of any energy store at all, or which allow one store to be set up, but dissipate the energy at some stage before ATP can be formed. In addition certain antibiotics are effective in blocking ATP formation without energy dissipation: oligomycin in mitochondria and Dio-9 in chloroplasts.

Attention has been concentrated on two hypotheses concerning the formation of the first energy store. In the chemical-intermediate version, a reduced redox carrier C combines with an unknown, I, and this complex (CH_2I) is oxidized forming $C \sim I$. In the chemiosmotic formulation, carriers of the type AH_2 are oxidized on one side of the thylakoid or crista membrane, releasing protons to a solution phase, and are re-reduced on the other side, taking up protons from the other solution phase. A protonmotive force (PMF) is thus set up. The protonmotive force can bring about the condensation of unknowns IOH and XH to form an intermediate $X \sim I$. Both hypotheses continue with the formation of ATP either by displacement of I by P_i and then X by ADP, or in reverse order, or by a concerted reaction.

The chemiosmotic hypothesis is often preferred in relation to the chloroplast since (i) it explains the formation of ATP following an acid–alkali transition, in the dark; (ii) it is not necessary to postulate specific intermediates, I, for each phosphorylation site; (iii) the primary energy store holds more high-energy bonds than can be accounted for by any known compound, but which is reasonable if due to a proton concentration difference between the inside and outside of the thylakoid; (iv) the action of the weak acid, fat-soluble type of primary uncoupling agent is explained; (v) Witt's group have observed pH changes of the order of magnitude required, and with the appropriate kinetics; (vi) phosphorylation always appears to require a closed vesicular system. None of this is however conclusive ground for eliminating the chemical-intermediate account.

The mechanism of phosphorylation appears to depend on coupling factors, which are 9 nm stalked spheres detachable from thylakoids and from cristae by EDTA and other media, and which are the site of oligomycin or Dio-9-sensitive ATPases. The ATPase of the chloroplast appears to be switched off by the simultaneous lack of ATP and light. Studies on exchange reactions offer inconclusive evidence on the order of addition of P_i and ADP during phosphorylation.

References

Arnon, D. I. (1967). In T. W. Goodwin (Ed.), *Biochemistry of Chloroplasts*, Vol. 2, Academic Press, London, p. 461.

Avron, M. and B. Chance (1966). In J. B. Thomas and J. C. Goedheer (Eds.), *Currents in Photosynthesis*, Donker, Rotterdam, p. 455.

Baltscheffsky, H. and L. V. von Stedingk (1966). In J. B. Thomas and J. C. Goedheer (Eds.), *Currents in Photosynthesis*, Donker, Rotterdam, p. 253.

Boyer, P. D. (1968). In T. P. Singer (Ed.), *Biological Oxidations*, Interscience, New York, p. 193.

Hind, G. and A. T. Jagendorf (1963). *Proc. Nat. Acad. Sci. U.S.*, **49**, 715.

Jagendorf, A. T. and J. Neumann (1965). *J. Biol. Chem.*, **240**, 3210.

Klingenberg, M. (1968). In T. P. Singer (Ed.), *Biological Oxidations*, Interscience, New York, p. 3.

Lardy, H. A. and S. M. Ferguson (1969). *Ann. Rev. Biochem.*, **38**, 991.

Livne, A. and E. Racker (1968). *Biochem. Biophys. Res. Commun.*, **32**, 1045.

Lynn, W. S., and K. D. Straub (1969a,b). *Proc. Nat. Acad. Sci. U.S.*, **63**, 540 and *Biochemistry*, **8**, 4789.

McCarty, R. E. and E. Racker (1967). *J. Biol. Chem.*, **242**, 3435.

Mitchell, P. (1966). *Biol. Revs.*, **41**, 445.

Racker, E. and T. E. Conover (1963). *Fed. Proc.*, Pt. 1, **22**, 1088.

Rumberg, B. and U. Siggel (1969). *Naturwissenschaften*, **56**, 130.

Witt, H. T., B. Rumberg and W. Junge (1969). In H. Staudinger and B. Hess (Eds.), *Biochemie des Sauerstoffs*, Mosbach symposium 1968, Springer, Berlin. p. 262.

CHAPTER 10

Chloroplast metabolism, and its relation to that of the cell

10.1 The chloroplast envelope

Chloroplasts are delimited by a double 'unit membrane' so that the envelope of the chloroplast is twice as thick as the comparable membranes found at the outer cell boundary and forming the endoplasmic reticulum. Although isolated chloroplasts have long been observed to have limiting membranes which can be distended and ruptured by osmotic shock, preparations of chloroplasts in which intact membranes were the rule rather than the exception were not reliably available until the necessary methods had been worked out by the group of Walker (see Walker, 1966). Since chloroplasts which have not leaked their ferredoxin can fix carbon dioxide in the light and evolve oxygen, an easy test for membrane damage is to observe this activity in the oxygen electrode. Furthermore, in the above system, when intact chloroplasts become depleted in carbon dioxide, oxygen evolution can be stimulated by the addition of certain intermediates of the reductive pentose cycle, but not others; this affords a test of the relative permeability of the chloroplast envelope to various compounds. Other methods depend on radio-carbon compounds, either by incubating isolated chloroplasts in a test-tube of medium, and analysing the pools inside and outside after illumination, or by using intact leaves and isolating the chloroplast at intervals after illumination (see Zalensky and Philippova, 1969). This last method makes use of the technique whereby the leaf is homogenized in a mixture of organic solvents, not miscible with water and of the appropriate density, so that the chloroplasts ('non-aqueous chloroplasts') are rapidly separated from the cytoplasm without being allowed to come into contact with an aqueous medium into which soluble metabolites might be lost.

From the first approach, it appears to be agreed that the envelope is much more permeable to phosphoglyceric acid, triosephosphates and pentose monophosphates than, say, to ribulose diphosphate, fructose-6-phosphate or sedoheptulose phosphate. From the second line of study comes the idea that equilibration of label between the chloroplast stroma and the cytoplasm may be a light-dependent process, being halted rapidly at the conclusion of the illumination of the leaf. It is of course possible that the activity of permeases (presumed to exist in the chloroplast envelope) and of enzymes at certain points in the metabolic pathways in the stroma are subject to control by the level of certain key metabolites in the stroma. Such control is a feature of pathways in the cytoplasm of, for example, mammalian liver cells and other tissues in which the classical studies of glycolysis and aerobic respiration were carried out. It is proper to note here that both the enzymes ribulose diphosphate carboxylase and fructose diphosphatase have activities much lower than is required to account for the observed rate of carbon dioxide fixation, assuming that their role according to the Calvin cycle is correct. Fructose diphosphatase has already been shown to lie at the heart of the control system for glycolysis and gluconeogenesis in animal tissues; with the accompanying enzyme phosphofructokinase, it would form a potent F6P-dependent ATPase if such control did not exist. With respect to the carboxylase, on the other hand, its great abundance has allowed several groups to achieve its purification ('fraction I protein'), but all efforts to detect activation effects have so far failed. While it may be true that the methods of preparation of this enzyme all result in an enormous degree of inactivation, there are at least two other views of the problem that deserve consideration. The first is that since fraction I protein is at least partly intimately associated with the thylakoid membranes, since carboxylation of RuDP is greatly slowed down in the dark, and since the half-saturating bicarbonate concentration in isolated chloroplasts is less than 6×10^{-4} M, as opposed to 2×10^{-2} M for the isolated carboxylase, it may be that the activity of the enzyme depends on its structural interaction with the thylakoid membrane, destroyed on isolation. Moreover, Bassham (1966) has laid some emphasis on the unequal labelling on carbons 3 and 4 in hexose, which suggests that the pools of dihydroxyacetone phosphate and phosphoglyceraldehyde are not in complete equilibrium (unlike the system in say muscle, where the enzyme aldolase which interconverts the two is one of the most abundant enzymes). It has been suggested that one of the two PGA molecules arising from the carboxylase reaction is retained on that enzyme so that it does not enter the triose phosphate pool. This remains only speculation. Still more speculative is the idea that there may be a direct reductive carboxylation forming one of the triose phosphates directly. Yet another point of interest is the relation of the carboxylase to the carbon dioxide incorporation pathway presented by Hatch and Slack.

10.2 The Hatch–Slack pathway

There is an important alternative to the RuDP–carboxylase system (proposed by Hatch and Slack, 1966) for the assimilation of carbon dioxide in certain plants. These authors* point out that certain tropical grass species are capable of incorporating carbon dioxide when the partial pressure of the gas is extremely low; the more usual pattern is for photosynthesis to decline and respiration to increase so that the concentration of carbon dioxide is never reduced below a certain point. These tropical grasses can pull the concentration of carbon dioxide down to virtually zero. They contain two characteristic enzymes in their chloroplasts: the first is pyruvate, inorganic phosphate, dikinase (an enzyme not previously known)

$$Pyruvate + P_i + ATP \rightarrow Phosphoenolpyruvate + AMP + PP_i$$

and the second phosphoenolpyruvate carboxylase (known to occur in plants)

$$PEP + CO_2 \rightarrow oxaloacetate + P_i \text{ (EC.4.1.1.31)}$$

These two enzymes together cause a rapid incorporation of carbon dioxide into oxaloacetate. Operation of NADP-linked malate dehydrogenase would convert oxaloacetate to malate with a favourable equilibrium constant. The carboxyl of the dicarboxylic acids is incorporated into sugar, probably RuDP as the acceptor. (Pyruvate must be regenerated for the reactions above.) They have shown that the RuDP-dependent light-driven carbon dioxide incorporation in chloroplasts from these species is much less than normal, and much less than the PEP-dependent fixation.

Johnson and Hatch showed that leaves in which the PEP-dependent carbon dioxide fixation took place were different in structure from the predominantly RuDP-dependent leaves. The difference lay in the vascular bundles of the leaves, which in the tropical-grass (PEP) group were surrounded by cylinders of bundle-sheath and mesophyll cells, causing the veins of the leaf to appear a darker green than the leaf lamina as a whole. In general, veins in leaves appear a lighter colour. The chloroplasts of the tropical grass group are dimorphic; the bundle sheath chloroplasts do not show well-developed grana, only roughly parallel thylakoids evenly spaced, with starch grains, while the mesophyll cells contain chloroplasts with grana, but without starch. The mesophyll chloroplasts are smaller than those of the bundle sheath. The presence of starch grains in only one type caused a density difference, which allowed a partial separation to be achieved by density-gradient centrifugation (Slack, Hatch and Goodchild, 1969). This in turn allowed the comparison of the enzymes present in the two types, and the results are shown in Table 10.1.

* The key observation is that, in sugar cane, labelled carbon dioxide first appears in the C_4-acids (Kortshak, Hartt and Burr, 1965).

Table 10.1. Distribution of enzymes in mesophyll and bundle-sheath chloroplasts in maize

Mesophyll	*Bundle-sheath*
Pyruvate, P_i dikinase	RuDP carboxylase
Malate dehydrogenase (NADP)	R5P isomerase
Glycerate kinase	Phosphoribulokinase
Nitrite reductase	FDP aldolase
Adenylate kinase	Alkaline FDPase
Pyrophosphatase	NADP-specific 'malic enzyme'
PEP carboxylase	

Common

Phosphoglycerate kinase, and NADP-specific G3P dehydrogenase

From the distribution of enzymes in Table 10.1 these authors concluded that the bundle-sheath chloroplasts operated cooperatively in synthesizing carbohydrate from precursors synthesized in the mesophyll from carbon dioxide by means of the C_4-dicarboxylic acid pathway. They drew attention to a remarkable development of plasmodesmata connecting the two types of cell, and other features of chloroplast ultrastructure which may be correlated with rapid transport of metabolites. At the time it was suggested that a C_4-dicarboxylic acid such as malate was formed in the mesophyll cells and transferred to the bundle sheath cells, where the —COOH group was transferred to RuDP by means of a transcarboxylase. Since however no transcarboxylase was discovered, and since the bundle sheath does possess an adequate quantity of RuDP carboxylase (Table 10.1), it is now agreed* to be more likely that the malate is decarboxylated in the bundle sheath, releasing CO_2 which is then fixed in the 'normal' manner by means of the reductive pentose cycle (Figure 10.1).

The advantage to the plant is that the C_4-acid pathway provides an ATP-operated CO_2-pump, which is able to raise CO_4 from a very low concentration in the mesophyll (too low for the operation of RuDP carboxylase) to a high level in the bundle sheath, even though an extra molecule of ATP is consumed per molecule of CO_2 fixed than most other plants.

This is very similar to the behaviour of the Crassulaceae which fix CO_2 at night, using respiratory ATP; they carry out photosynthetic re-fixation of the CO_2 during the day, when their stomata are closed to prevent the loss of water. This is known as *Crassulacean Acid Metabolism*.

Laetsch (1969) has described the structure of the bundle-sheath chloroplasts in several monocotyledenous species, and considers that the tropical grasses (including sugar-cane and maize) show an evolution of bundle-sheath chloroplasts into amyloplasts. He also commented on similarities in the dicotyledon genera *Amaranthus*

* The reader's attention is directed to a symposium recently held in Canberra on both this subject and photorespiration to be published by John Wiley and Sons in 1971.

and *Atriplex*. This in fact is part of a classification, in part worked out by Prat (1936), which has been extended in a paper by Tregunna, Downton and Jollife (1969). They compared leaf anatomy as above with the compensation point of carbon dioxide concentration. Above that critical concentration, the rate of carbon dioxide uptake by photosynthesis exceeds its output by respiration, but at lower concentrations the position is reversed and there is a net loss of carbon by the plant. It happens that most plants can be classed as L for low compensation (5 ppm

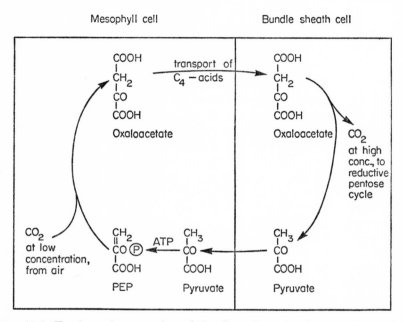

Figure 10.1. To show the operation of the C_4-pathway in the mesophyll cells of sugar cane acting as a CO_2-pump, feeding the reductive pentose pathway in the bundle-sheath cells

CO_2) or H for high (35–50 ppm). The H group increases its net rate of photosynthesis when the oxygen concentration is reduced from 21 to 3 %, while the L group are hardly affected. This last effect is probably a manifestation of the phenomenon of photorespiration (see the next section).

10.3 Photorespiration

The earlier measurements of the rate of photosynthesis by algal cells, including the classical studies of Emerson and Warburg were subject to an unavoidable assumption that the rate of respiratory oxygen uptake and carbon dioxide evolution, measured accurately enough in the dark, remained constant during illumination. This rate was subtracted from the observed photosynthetic gas exchange, giving the 'rate of photosynthesis'. This was serious enough ; Emerson regarded the low quantum requirement measured by Warburg as an error stemming in part from the very high respiratory rates of his algae under the experimental conditions (see Emerson, 1949). Still more serious was the suspicion that the rate of respiratory gas exchange might not be constant. With the use of isotopes this suspicion became a fact. Isotopes allowed the rate of respiration of say ^{14}C sugars to be measured independently from the rate of photoassimilation of say $^{12}CO_2$, and similarly with oxygen exchanges using say $^{18}O_2$ in the gas phase, and $^{16}O_2$ elsewhere. The topic of photorespiration may be said to cover all effects where the assumption of the constancy of respiratory gas exchange was found not to hold.

Two effects may be briefly mentioned. The first is the concept of the chloroplast having respiratory activity. This was for a time an objection to Arnon's concept of photophosphorylation, and was settled by the demonstration that isolated chloroplasts, both intact and broken, showed none of the pathways for respiratory metabolism or oxidative phosphorylation. It remains true that the chloroplasts of many species of plant contain enzymes such as polyphenol oxidase, and there was a period when it was thought that such oxidases might provide a terminal pathway for the oxidation of NADH and NADPH, generated by respiratory metabolism. However, although the role of these enzymes is not at all clear, there is no basis whatever for holding that the chloroplast has any respiratory activity.

The second effect is a somewhat vague but well known effect of blue light (420–480 nm) on the carbon metabolism of plants, which may possibly operate by stimulating the respiration system. The effect is independent of chlorophyll, and saturates at a very low intensity of light. Moreover, after blue-light treatment cells take several hours to recover their normal state. The location in the cell of this effect is not known, although Ogasawara and Miyachi (1969) find that there is apparently an increase in the flow of carbon through tricarboxylic acid cycle intermediates. It has been tentatively suggested that this blue-light effect is a carotenoid-mediated process. The reader is referred to the above paper, and to a brief cryptic review by French (1966) for further information on this curious subject.

Having stressed that isolated chloroplasts (from higher plants) do not respire, we are faced by the observation that under certain conditions the rate of respiration of some algae and higher plants increases markedly in the light by 2–3 times. Furthermore, Bidwell, Levin and Shepard (1969) showed that the carbon dioxide released by this increased respiration contains a high proportion of carbon recently fixed by photosynthesis; the newly

7

fixed carbon is available for respiration before it has time to come into equilibrium with the carbohydrate substrates of the cytoplasm of the cell. Referring back now to section 10.1, we find that Zalensky and Philippova (1969) observed a rapid release of fixed carbon dioxide into the cell cytoplasm, which was quickly suppressed when the illumination ceased. If we put these observations together, it may be inferred that during illumination newly-fixed carbon permeates through the chloroplast envelope and gives rise to carbon dioxide. This may be produced by mitochondrial respiration of the usual respiratory substrates, such as triose phosphate, but it is unlikely since the effect is stimulated by cyanide (see below) and by increasing the oxygen concentration above the value in air. Mitochondrial respiration in most plants relies mainly on cytochrome oxidase which is inhibited by cyanide, and which has such a high affinity for oxygen that the rate is not dependent on the concentration over a wide range. However in some plant mitochondria such as skunk cabbage (*Symplocarpus foetidus*) and *Arum* spadices there is an alternative oxidase that provides a fast, cyanide-stable, respiration pathway.

A more likely possibility is that photorespiration is based on the production of glycolate in the chloroplast and its subsequent oxidation in the cytoplasm. Zelitch (1959) showed that leaves such as tobacco formed glycolate at a high rate and this was indeed oxidized by glycolate oxidase, which appears to be largely if not entirely present in the cell cytoplasm. Cell organelles such as peroxisomes (microbodies) (see Plate 1) are a likely site for this oxidation. We shall discuss the proposed 'glycolate pathway' in the next section; it will be seen that the oxidation of glycolate is necessary for some biochemical syntheses.

In the '*Warburg effect*' it is observed that photosynthesis is inhibited by high concentrations of oxygen. Much of this inhibition may be due to a sharp increase in the rate of photorespiration by means of the glycolate pathway. (As mentioned in Chapter 8, oxygen also tends to oxidize Q, and might lead to an inhibition of the electron transport pathway. It is worth noting that the concentration of oxygen in the gas phase affects mainly the cytoplasm of the cell in the light; the chloroplast produces oxygen which must presumably mean that the concentration is always locally high.)

At low concentrations of carbon dioxide photosynthetic carbon dioxide fixation diminishes (the RuDP-dependent process being more susceptible than the PEP-dependent mechanism) while photorespiration continues at a rate some two times as great as the dark respiration. (The production of glycolate is known to be accelerated by low carbon dioxide concentrations.) There is a general correlation between the presence of the Hatch–Slack pathway and the absence of photorespiration.

The above considerations also provide an explanation for the Kok effect, which was the observation that the quantum requirement for photosynthetic gas exchanges was higher below the compensation point of light intensity, (that is, in dim light where there was a net uptake of oxygen by respiration) than above it (see Kok, 1951). The isotope study by Hoch and Owens (1966) showed that above a certain level, increasing the light intensity (particularly system II light) caused both increased photosynthesis and increased respiratory exchange. These effects are reproduced in Figure 10.2. Hoch and Owens also showed that the effect of cyanide, shown by Warburg (1920) to inhibit photosynthesis only at high light intensities, was due not to any inhibition of photosynthesis proper but to a stimulation of the photorespiration process.

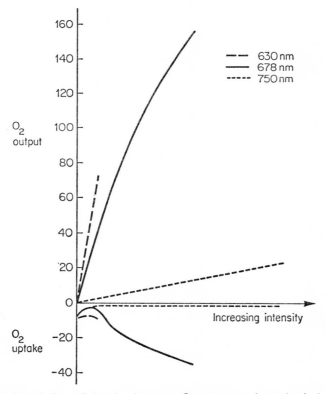

Figure 10.2. Resolution of the simultaneous O_2 output and uptake (using isotopic oxygen), and their dependence on the wavelength and intensity of light. From G. Hoch and O. v. H. Owens (1963). In *Photosynthetic Mechanisms of Green Plants*, Publ. No. 1145, NAS-NRC, Washington, D.C., p. 409, with permission

10.4 The glycolate pathway

As a matter of convenience, we discuss most topics of photosynthesis in green plants in terms of the equation of de Saussure, with hexose as the product.

$$6CO_2 + 6H_2O \rightarrow C_6H_{12}O_6 + 6O_2$$

Nevertheless depending on the conditions of growth algae may present protein, fat or acids such as glycolic and malic acids as the principal product

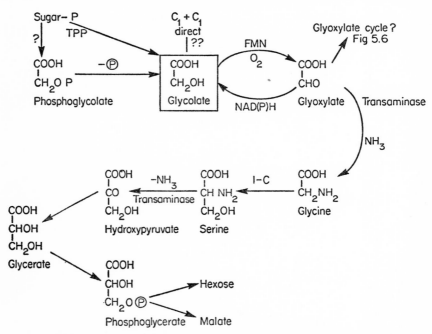

Figure 10.3. A working hypothesis for the 'glycolate pathway'.

of photosynthesis. It is possible to set out a tentative scheme to account for these products, and this area of metabolism may be referred to as the glycolate pathway (Figure 10.3).

Glyoxylate and glycolate are formed particularly when there is a high concentration of oxygen, and relatively little carbon dioxide. The use of tracers indicates that incorporation of carbon dioxide into the two-carbon compounds proceeds via PGA; also the two carbon atoms are equally labelled from $^{14}CO_2$ (Tolbert, 1963). He found evidence that phosphoglycolate was formed first, then hydrolysed by a vigorous phosphatase specific to chloroplasts. Bassham (1963), suggesting that the two-carbon units arose from the thiamine pyrophosphate–glycolaldehyde adduct which is implicated in the transketolase reactions of the Calvin cycle, admitted that while such cleavage was known in bacteria (the phosphoroclastic split) it had not been demonstrated in the chloroplast. The suggestion is interesting, however, since the phosphoroclastic split, resulting in acetyl phosphate, would provide a source for the synthesis of fatty acids via acetyl CoA.

In the reactions shown in Figure 10.3, serine can occur either from PGA, when it is mainly carboxyl labelled, or it can be uniformly labelled; such uniform labelling has been taken to indicate formation from glycine, uniformly labelled from glyoxylate, which can form serine by tetrafolic acid-linked reactions.

$$\text{Glyoxylate} \xrightarrow{\text{THFA}} \text{1-C} + CO_2 \text{ (possibly via oxalate and formate)}$$

$$\text{Glycine} + \text{1-C} \rightarrow \text{Serine (EC.2.1.2.1.)}$$

Malate could be formed in theory by condensation of glyoxylate with acetyl-CoA, as in the glyoxylate pathway of certain germinating seeds (*Ricinus*) and bacteria. (The enzyme has not yet been demonstrated in chloroplasts, however.) This source of malate is different from that of the Hatch–Slack pathway (section 10.2).

10.5 The autonomy of the chloroplast

While it is clear that the chloroplast depends on its cytoplasmic surroundings for water, carbon dioxide and the pool of metabolic substrates, it has features which suggest that within the double envelope, the chloroplast duplicates in several features of the whole cell.

First, the chloroplast, like the nucleus, appears to have its own pool of ATP and ADP which do not exchange with the cytoplasmic pools. Secondly, again like the nucleus and mitochondrion, there DNA and RNA are present, some of the latter being in the form of ribosomes. In the nucleus, DNA is the genetic material, consisting of a linear sequence of bases, some millions in total but of only four types. Sections of the sequence are transcribed by an enzyme (DNA-dependent RNA polymerase), resulting in the formation of RNA (*messenger*). The ribosome, with messenger RNA attached, in some way assembles amino acids (each activated by a specific *transfer* RNA) so that a chain of amino acids is formed, the sequence being governed by the sequence of the bases in the messenger and hence of the corresponding section of the DNA. Each chain of amino acids constitutes a protein molecule, and since enzymes are proteins, the mechanism outlined above provides for the synthesis of enzymes, their composition governed by the sequence of bases in the DNA of the nucleus.

The ribosomes of the chloroplast are often seen in groups, suggesting that they are attached to a messenger ('polyribosomes'), and there is evidence that chloroplasts can carry out incorporation of amino acids into protein. However an important question is the source of the messenger. If it were transcribed from the chloroplast DNA (the chloroplast RNA polymerase does occur) then it would be supposed that mutations arising in the chloroplast DNA should result in chloroplast characteristics showing non-Mendelian inheritance. On the one hand, there are plastid characteristics which are inherited in a non-Mendelian fashion, but on the other hand mutations which result in deficiencies which can be shown to result in alterations to known proteins (many of Levine's mutant algae are of this type) have usually been found to follow Mendelian genetics. That is, the proteins are under the control of the nuclear DNA. This does not of course establish that the protein was coded by the sequence of nuclear DNA; DNA has other roles which include the regulation of which protein groups are to be transcribed. It is quite conceivable that nuclear DNA should in some way suppress (or fail to activate) the transcription of chloroplast DNA for synthesis of chloroplast proteins in these cases. We need mutations that allow the protein to be recognized, but which give it some new property due to a slightly altered amino acid sequence. (This was the case in Ingraham's classic study of 'sickle-cell' haemoglobin.) Mendelian inheritance of such mutations would indicate that the nuclear DNA carried the coding for those proteins; non-Mendelian inheritance would establish the role of chloroplast DNA.

For a very comprehensive survey of the genetic material of the plastid the reader is referred to Chapter X of the work by Kirk and Tilney-Bassett (1967).

References

Bassham, J. A. (1963). In *Photosynthetic Mechanisms of Green Plants*, publication 1145, NAS-NRC, Washington D.C., p. 635.

Bidwell, R. G. S., W. B. Levin and D. C. Shepard (1969). *Plant Physiol.*, **44**, 946.

Emerson, R. and M. S. Nishimura (1949). In J. Franck and W. E. Loomis (Eds.), *Photosynthesis in Plants*, Iowa State College Press, Ames, Iowa, p. 219.

French, C. S. (1966). In T. W. Goodwin (Ed.), *Biochemistry of Chloroplasts*, Vol. 1, Academic Press, London, p. 377.

Hatch, M. D. and C. R. Slack (1966). *Biochem. J.*, **101**, 103.

Hoch, G. and O. v. H. Owens (1963). In *Photosynthetic Mechanisms of Green Plants*, publication 1145, NAS-NRC, Washington, D.C., p. 409.

Kirk, J. T. O. and R. A. E. Tilney-Bassett (1967). *The Plastids, their Chemistry, Structure, Growth and Inheritance.* Freeman, London. pp. 302, 338, 574.

Kok, B. (1951). In *Carbon Dioxide Fixation and Photosynthesis, Symp. Soc. Exp. Biol.*, Vol. 5, Cambridge University Press, p. 211.

Kortshak, H. P., C. E. Hartt and G. O. Burr (1965). *Plant Physiol.* **40**, 209.

Laetsch, W. M. (1969). In H. Metzner (Ed.), *Progress in Photosynthesis Research*, Vol. 1, Institut fur Chemische Pflanzenphysiologie, Tubingen, p. 36.

Ogasawara, N. and S. Miyachi (1969). In H. Metzner (Ed.), *Progress in Photosynthesis Research*, Vol. 3, Institut fur Chemische Pflanzenphysiologie, Tubingen, p. 1653.

Philippova, L. A. and O. V. Zalensky (1969). *Photosynthetica*, **3**, 104.

Prat, H. (1936). *Ann. Sci. Nat. (Botan.) Ser*, **10**, 18, 165.

Slack, C. R., M. D. Hatch and D. J. Goodchild (1969). *Biochem. J.*, **114**, 489.

Tolbert, N. E. (1963). In *Photosynthetic Mechanisms of Green Plants*, publication 1145, NAS-NRC, Washington D.C., p. 648.

Tregunna, E. B., J. Downton, and P. Jolliffe (1969). In H. Metzner (Ed.), *Progress in Photosynthesis Research*, Vol. 2, Institut fur Chemische Pflanzenphysiologie, Tubingen, p. 488.

Walker, D. A. (1967). In T. W. Goodwin (Ed.), *Biochemistry of Chloroplasts*, Vol. 2, Academic Press, London, p. 53.

Warburg, O. (1920). *Biochem. Z.*, **103**, 188.

Zelitch, I. (1959). *J Biol. Chem.*, **234**, 3077.

Prospect

If a summary of the foregoing summary treatment can be permitted, I should stress the 'approximate' nature of the account in these terms:

Photosynthesis in green plants takes place in chloroplasts whose relationship with their surrounding cells is approximately understood; the process is reasonably well apportioned into a metabolic pathway, of which perhaps all but a few details can be confidently set out, and a membrane-bound portion. The membranes or thylakoids contain an electron transport system concerning which there is considerable information (notwithstanding the considerable room for argument) and a phosphorylation process which can hardly escape elucidation relatively soon. The inscrutability of the oxygen-evolving reaction has begun to crack, and the physicists' attack on the light-harvesting and photochemical apparatus is maintaining its momentum. Finer and finer models are prepared each year to display the increasing quantity of data concerning membrane structure itself, and in all these parts of the photosynthetic topic we find investigators from all walks of biochemistry converging, bring complementary ideas on membrane structure, metabolism, phosphorylation, photobiology, electron transport and electron microscopy. From this viewpoint of approximate optimism, what do we see in the future? Will the 1984 textbooks be Final Editions? Perhaps all the pieces of the jigsaw will fit, or be declared to fit, and there will be a scholastic exodus on the scale of that of 1453. On the other hand, perhaps the biochemistry of photosynthesis is reaching a point like that reached by physics at the turn of the century, ready for an Einstein to demonstrate the gaps in the jigsaw to be more significant than the areas of apparent contact. No prediction will be made here; fortune-tellers are notoriously vulnerable to the unexpected. Personally, however, I find it exciting to envisage photosynthesis as a focal point for a new modern synthesis of biology.

Appendix: physical constants, formulae, etc.

	symbol	value
Velocity of light *in vacuo*	c	2.998×10^8 m sec^{-1}
Planck's constant	h	6.626×10^{-27} erg sec
		6.626×10^{-34} J sec
electronic charge	e	4.803×10^{-10} e.s.u.
		1.602×10^{-19} C
electron-volt (unit of energy)	eV	1.602×10^{-19} J
		equivalent to 23.053 cal mol^{-1}
		equivalent to photon of
		1239.5 nm wavelength
calorie (defined)	cal	4.1841 J
Joule (absolute)	J	1.0 CV $= 10^{-7}$ erg
		$= 0.2390$ cal
Avogadro's Number		
(molecules per mole)	N	6.0225×10^{23}
Absolute zero	$0°$K	$-273.16°$C
Faraday constant	F	96 649 C equiv.$^{-1}$
Gas constant	R	8.3143 J deg.$^{-1}$ mol^{-1}
		$= 1.987$ cal deg.$^{-1}$ mol^{-1}
		$= 0.08206$ litre atm. deg.$^{-1}$ mol^{-1}

Solar constant (radiant energy entering the earth's atmosphere) 2.0 cal min^{-1} cm^{-2}

$RT \log_e = 5804 \log_{10}$ J mol^{-1} at $25°$; *add* 19 J per degree rise in temperature.
  $= 1387 \log_{10}$ cal mol^{-1} at $25°$; *add* 4.6 cal per degree rise in temperature.
$(RT/nF) \log_e = (59.16/n) \log_{10}$ mV at $25°$; *add* 0.20 mV per degree rise in temperature.

Wavelengths of hydrogen lines used in the calibration of spectrophotometers:

$$656.3 \text{ nm, } 486.1 \text{ nm.}$$

Equations for the estimation of chlorophyll extracted in 80% acetone:

$$C_a = 12.7 E^{1cm}_{663nm} - 2.69 E^{1cm}_{645nm}$$

$$C_b = 22.9 E^{1cm}_{645nm} - 4.68 E^{1cm}_{663nm}$$

From Hill, R. (1963). In *Comprehensive Biochemistry*, M. Florkin and E. H. Stotz, (Eds.), Vol. 9, Elsevier, Amsterdam, Chapter 3, p. 73. C_a and C_b are the concentrations of chlorophylls a and b respectively in mg/litre; E^{1cm}_{663nm} is the extinction (absorbance) ($\log I_0/I$) of a 1 cm path-length sample at a wavelength of 663 nm.

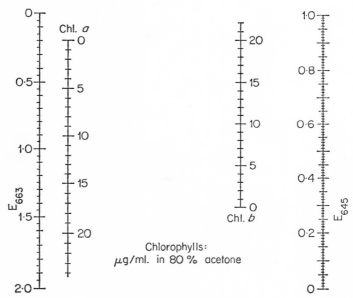

Figure A.1

Figure A.1 presents a nomogram for the rapid solution of the above equations: locate the extinction values on the appropriate axes (note that the directions of the scales are reversed) and join by a thread or a transparent ruler (but beware of refraction effects). The values of C_a and C_b may be read off from the points of intersection of the thread with the appropriate scales.

Answers to numerical problems

1. Once, on average, in 0·25 sec.
   One in $3·6 \times 10^8$ molecules.

2. (i) That the photosynthetic unit was of variable size (since the means of the groups were so far apart in terms of their standard deviations). The sizes were 272, 630, 1130, 2550, 5100 chlorophyll molecules per unit, respectively. These lie close to a doubling series (1, 2, 4, 8, 16), and suggest that larger units are formed as aggregates of small ones.
   (ii) In the mutant the same groups appear (with barely significant differences) but whereas the normal leaf had a preponderance of the 'classical' size unit (2550) the mutant showed the most common size to be 602.
   (iii) That the size of the photosynthetic unit may be under the control of developmental or photosynthetic factors. See 'plastic chlorophyll', pp. 96 and 114.
   Data composed from published work of Schmid, G. H. and Gaffron, H., (1968). *J. Gen. Physiol.* **52**, 212.

3. The energy of activation is 1·0 eV, or 23 000 cal. mol^{-1}. The energy of a 690 nm photon is 1·8 eV, therefore the store has an energy level of 0·8 eV (which is similar to the potential difference between X_{II} and Y_{II}, see section 8.2).
   Data composed from work of Bertsch and Azzi, cited in Bertsch, W. F., (1969). In H. Metzner (Ed.), *Progress in Photosynthesis Research*, Vol. 2, Institut für Chemische Pflanzenphysiologie, Tubingen, p. 996.

4. (a) 16·56 kcal mol^{-1}.
   (b) 0·36 V if 1 ATP was formed for 2 electrons passing.
   (c) Between water/oxygen ($E_0' = +815$ mV at pH 7) and Y_{II} (unknown); between PQA ($E_0 = 115$ mV *in ethanol*) and ferricyanide (if only System II is operating if system I is included, ATP could be coupled to the span X_I to ferricyanide, or to PQA to P700). See section 8.2. *Problem by courtesy of Dr. A. R. Crofts.*

5. The effect of uncoupling agents was presumably to interfere with the formation of either a high-energy compound or a field of some kind. By plotting the absorbance change against log $[K^+]_{outside}$, it is clear that the shift is proportional to the calculated membrane potential. The extent of the shift observed with

185

illumination is (by extrapolation) found to be equivalent to $5 \cdot 62 \times 10^4$ M K$^+$, giving a value from the equation for $\psi = 287$ mV. This is a reasonable value for phosphorylation, particularly if a pH gradient is also present. See Chapter 9. *Problem by courtesy of Dr. A. R. Crofts.*

6. (i) In the case of hexose, derived from F6P, the label initially is predicted to be located in carbons 3 and 4, which should be equal in activity. During subsequent turns the other atoms become labelled, but C3 and C4 remain the most active, and equally so.

   (ii) The experimental data are in accord with the prediction that C3 and C4 would be the most active, but there is a slight but noticeable difference between the activities of C3 and C4 that requires additional explanation. This is still an open matter, but possibilities are perhaps that one of the PGA molecules is retained and reduced by a different mechanism from the other, so that the pools of DHAP and G3P are not properly equilibrated, or, perhaps, the operation of the glycolate pathway (see section 10.4) could upset the symmetry of the transketolase reactions.

   Problem composed from data of Gibbs, M., (1963). In *Photosynthetic Mechanisms of Green Plants*, publication no. 1145, NAS/NRC, Washington, D.C. p. 663.

7. 2×10^9.

8. 4210.

9. Three orders of magnitude (10^3).

Index